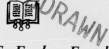

Once
Upon
a Town

BOOKS BY BOB GREENE

Once Upon a Town

Duty: A Father, His Son, and the Man Who Won the War

Chevrolet Summers, Dairy Queen Nights

The 50-Year Dash

Rebound: The Odyssey of Michael Jordan

All Summer Long

Hang Time

He Was a Midwestern Boy on His Own

Homecoming: When the Soldiers Returned from Vietnam

Be True to Your School

Cheeseburgers

Good Morning, Merry Sunshine

American Beat

Bagtime (with Paul Galloway)

Johnny Deadline, Reporter

Billion Dollar Baby

Running: A Nixon McGovern Campaign Journal

We Didn't Have None of Them Fat Funky Angels on the Wall of Heartbreak Hotel, and Other Reports from America

BOOKS BY BOB GREENE AND D. G. FULFORD

To Our Children's Children Journal of Family Memories

Notes on the Kitchen Table: Families Offer Messages of Hope for Generations to Come

To Our Children's Children: Preserving Family Histories for Generations to Come

Once Upon a Town

The Miracle of the North Platte Canteen

BOB GREENE

WILLIAM MORROW
An Imprint of HarperCollins*Publishers*

Grateful acknowledgment is made for permission granted by the author to quote a passage from *All Aboard!: The Railroad in American Life,* by George H. Douglas, © 1992 George H. Douglas.

HarperCollins books may be purchased for educational, business, or sales promotional use. For information, please write to Special Markets Department, HarperCollins Publishers Inc., 10 East 53rd Street, New York, NY 10022.

FIRST EDITION

Designed by Paula Russell Szafranski

Printed on acid-free paper

Library of Congress Cataloging-in-Publication Data has been applied for.

ISBN 0-06-008196-1

02 03 04 05 06 WBC/QW 10 9 8 7

For Keith and Mary Ann Blackledge

One

On Interstate 80, three or four hours into the long west-ward drive across Nebraska, with the sun hovering merci-lessly in the midsummer sky on a cloudless and broiling July afternoon, there were moments when I thought there was no way I'd ever find what I had come here to seek:

The best America there ever was. Or at least whatever might be left of it.

It wasn't some vague and gauzy concept I was searching for; not some version of hit-the-highway-and-aimlessly-look-for-the-heart-of-the-nation. This was specific: a real town.

But the news, as I was hearing it from the rental-car

radio on this particular summer's day, made Nebraska in the early years of the twenty-first century sound deflatingly like the rest of the continental United States.

In Sutherland—not far from where I was heading—a man had come home from work to the rural farmhouse he and his sixty-six-year-old wife shared. The house, located on a dirt road about a mile from the closest neighbor, was in an area so quiet and sedate that there was seldom a reason to lock the doors. When the man arrived home, he found his wife sitting in a chair dead, with a gunshot wound to her head.

Two men—Billy J. Reed, twenty, and Steven J. Justice, twenty-two—were soon arrested. Prosecutors said they were wanted for the recent murders of an elderly couple in Adams County, Illinois. The men allegedly were fleeing across Nebraska, and stopped in at the farmhouse in Sutherland with the intention of robbing it. The men evidently selected the farmhouse at random, and allegedly shot the sixty-six-year-old woman to death just because she happened to be at home.

Also in Nebraska on this summer day, Richard Cook, thirty-four, was sentenced to life in prison because of what he did to a nineteen-year-old woman who was a college freshman.

She had been driving late at night when her car suffered a flat tire. Alone, she had pulled over to the side of the road to try to change the tire. Richard Cook, driving on the same road that night, stopped his car as if to help the stranded young woman. He then assaulted her, shot her five times, and dumped her body in the Elkhorn River.

In Hall County, a man named Jamie G. Henry, twenty-four, was under arrest for allegedly using an electrified cattle prod to discipline his eight-year-old stepson. The cattle prod, according to sheriff's deputies, was of the kind designed to jolt two-thousand-pound bulls into obedience. Jamie Henry reportedly used it on the boy and his five-year-old sister; Henry also allegedly punished the boy by tying him tightly at his hands and ankles, and, during the winter, tying the boy barefoot to a tree and locking him out of the house in the cold.

That is what was going on in Nebraska on this summer day—at least that is what was going on that had been deemed worthy of the public's notice. It could have been anywhere in the United States; the police-blotter barbarism of the news, the seeming soullessness of the crimes, had a sorrowful and deadening familiarity to them.

Yet once upon a time, in the town I hoped to reach by nightfall . . .

Well, that was the purpose of this trip. Once upon a time—not really so very long ago—something happened in this one little town that, especially on days like this one, now sounds just about impossible. Something happened, in the remote Nebraska sandhills, in a place few people today ever pass through. . . .

Something happened that has been all but forgotten. What happened in that town speaks of an America that we once truly had—or at least that our parents did, and their parents before them.

We're always talking about what it is that we want the country to become, about how we can save ourselves as a people. We speak as if the elusive answer is out there in the mists, off in the indeterminate future, waiting to be magically discovered, like a new constellation, and plucked from the surrounding stars.

But maybe the answer is not somewhere out in the future distance; maybe the answer is one we already had, but somehow threw away. Maybe, as we as a nation try to make things better, the answer is hidden off somewhere, locked in storage, waiting to be retrieved.

That's what I was looking for on this Nebraska summer afternoon, with the temperatures nearing one hundred degrees. The car radio continued to tell the dismal breaking news of the day, and I continued on toward my destination, a town with the unremarkable name of North Platte.

Two

North Platte, Nebraska, is about as isolated as a small town can conceivably be. It's in the middle of the middle of the country, alone out on the plains; it is hours by car even from the cities of Omaha and Lincoln. Few people venture there unless they live there, or have family there.

But before the air age, the Union Pacific Railroad's main line ran right through North Platte. In 1941, the town had little more than twelve thousand residents. When World War II began, with young men being transported across the American continent to both coasts before being shipped out to Europe and the Pacific, those Union

Pacific cars carried a most precious cargo: the boys of the United States, on their way to battle.

The trains rolled into North Platte day and night. A local resident—or so I had heard—came up with an idea:

Why not meet the trains coming through, to offer the servicemen a little affection and support? The soldiers were out there on the empty expanses of midwestern prairie, filled with thoughts of loneliness and fear. Why not try to provide them with warmth and the feeling of being loved?

On Christmas Day 1941, it began. A troop train rolled in—and the surprised soldiers on board were greeted by North Platte residents with welcoming words, heartfelt smiles and baskets of food and treats.

What happened in the years that followed was nothing short of amazing—some would say a miracle. The railroad depot on Front Street was turned into the North Platte Canteen. Every day of the year—from 5 A.M. until the last troop train of the night had passed through after midnight—the Canteen was open. The troop trains were scheduled to stop in North Platte for only ten minutes at a time before resuming their journey. The people of North Platte made those ten minutes count.

Gradually, word of what was happening in North Platte spread from serviceman to serviceman during the war, and on the long train rides across the country the soldiers came to know that, out there on the Nebraska flatlands, the North Platte Canteen was waiting for them.

Each day of the war—every day of the war—an average of three thousand to five thousand military personnel came through North Platte, and were welcomed to the Canteen. Toward the end of the war, that number grew to eight thousand a day, on as many as twenty-three separate troop trains.

Many of the soldiers were really just teenagers. This was their first time away from home, the first time away from their families. On the troop trains they were lonesome and far from everything familiar, and they knew that some of them might never come back from the war, might never see their country again. And then, when they likely felt they were out in the middle of nowhere, they rolled into a train station and were greeted day and night by men, women and children who were telling them thank you, were telling them that their country cared about them.

The numbers are almost enough to make you cry. Remember—only twelve thousand people lived in that secluded town. But during the war, six million soldiers passed through North Platte, and were greeted at the train station that had been turned into a Canteen. This was not something orchestrated by the government; this was not paid for with public money. All the food, all the services, all the hours of work were volunteered by private citizens and local businesses.

The only federal funding for the North Platte Canteen was a five-dollar bill that President Roosevelt sent from the White House because he had heard about what was taking place in North Platte, and he wanted to help.

It might have been a dream—but it wasn't. Six million soldiers who passed through that little town—six million of our fathers, before we were born. And every single train was greeted; every man was welcomed.

It was a love story—a love story between a country and its sons.

And it's long gone.

That is why I was traveling across Nebraska on this sun-baked July afternoon.

There is no reason for anyone to pass through North Platte anymore—the jet age has done away with that. If a person wants to get from one end of the United States to the other, he or she now likely does it five miles in the air, high above the country—high above Nebraska. All the small towns flash by in an instant—on a cloudy day, it's as if they are not even down there.

And the country itself . . . the country itself at times seems to have gone away. At least a country in which neighbors would join together for five straight years, every day and every night, just so they could provide kindness and companionship to people they had never met.

In a lot of ways, it is a country that many of us seem always to be searching for.

I wasn't at all certain what I would find when I got to North Platte.

But the people from the Canteen—the people who came there on their own time to run it, the people who hurriedly ran inside to savor it, on their way to war—will soon all be gone.

I wanted to get to North Platte before it was too late.

Three

The Beatles and the Goo Goo Dolls sang consecutive songs on the car radio. The interstate stretched the breadth of Nebraska; the air conditioner in the rental car from the Omaha airport blasted coolly from the dashboard, and with the windows rolled up, the farms and ranches and small towns might have been postcards instead of real life.

The music played all the way across the state—as soon as one town's station faded out, another would drift in. The world outside the car seemed mute, locked out by the music.

I thought about the six million soldiers, each inside his own world as they moved across Nebraska confined to the railroad cars in which they rode. No interstates in the 1940s; no air-conditioning on the trains. The Nebraska outside the soldiers' train-compartment windows had to have seemed very close to them indeed—no locking that Nebraska out.

What must they have been thinking, on their way across? I knew my destination—North Platte. They didn't know theirs—not precisely. They didn't even know if they would ever come home again.

Every few miles, to the side of the road, I would see a blue sign with a white star pattern. The sign informed motorists that we were on the Eisenhower Interstate System. He had been the one who started to build the interstates—after he was back from Europe, living in the White House.

In the years the soldiers rolled through these same plains, the years before Ike's superhighways, he was somewhere else. Across an ocean, he waited for them.

They must have believed that no one even knew they were here.

So much emptiness. Everything, it seemed, was off the main road—I stopped for a sandwich in Grand Island, halfway across, and it was a good seven miles from the highway into town.

They had to have seen an occasional farmhouse. But

especially at night, especially in the blackness, they could have been excused for thinking they were moving through the plains like apparitions, in secret.

Near the town of Gothenburg, I saw a placard that said it used to be a Pony Express stop. The Iron Horse, in the 1940s, had to have seemed quite modern, compared to that. At least the trains were engine-driven—an improvement over the ponies. The trains evidently were quite efficient, moving their passengers toward the war.

As I passed Lexington and headed in the direction of Cozad, I could see, for the first time, the tracks. A freight train was rolling in the same direction I was. The tracks would carry it into North Platte. The train was off to the north, and soon I lost sight of it.

At the North Platte exit I checked into the Quality Inn, a minute off the highway. You could sleep here, rest for the night, have breakfast, and be back on the interstate without ever seeing the town itself.

But I would be staying for a while.

The idea for the Canteen, it turned out, was the offspring of a mistake.

Ten days after Pearl Harbor, the families and friends of members of the Nebraska National Guard's Company D heard a rumor: Their sons, buddies and sweethearts would be coming through North Platte on a troop train on their way to the West Coast. Military movements were confidential. But even with no announcement, about five hundred of the townspeople came to the station with food, cigarettes, letters and love to give to the boys.

The train finally arrived. The people of North Platte hurried toward the cars.

But the soldiers on board were not Company D of the Nebraska National Guard—they were Company D of the Kansas National Guard.

After an awkward few moments, the North Platte residents began to pass out their gifts to the soldiers from Kansas. These hadn't been the boys the townspeople had been waiting for—the boys the townspeople knew—but it wasn't the soldiers' fault. The men, women and children of North Platte wished the Kansas soldiers the best of fortune, made certain they had all the presents that had been intended for the Nebraska troops, and waved them on their way.

A woman named Rae Wilson—twenty-six years old, a store clerk in town, the sister of the young commander of Nebraska Company D—wrote a letter to the North Platte *Daily Bulletin,* a newspaper that is now dead.

The brief article containing her letter, from the edition of December 18, 1941, is still on file in town:

Following the visit of the troop train here yesterday afternoon, Miss Rae Wilson, sister of North Platte's Captain Denver Wilson, suggested that a canteen be opened here to make the trips of soldiers thru the city more entertaining. She offered her services without charge. Her public-spirited and generous offer is contained in the following communication to the *Bulletin:*

Editor, The Daily Bulletin:
I don't know just how many people went to meet the trains when the troops went thru our city Wednesday, but those who didn't should have.

To see the spirits and the high morale among those soldiers should certainly put some of us on our feet and make us realize we are really at war. We should help keep this soldier morale at its highest peak. We can do our part.

During World War I the army and navy mothers, or should I say the war mothers, had canteens at our own depot. Why can't we, the people of North Platte and the other towns surrounding our community, start a fund and open a Canteen now? I would be more than willing to give my time without charge and run this canteen.

We who met this troop train which arrived about 5 o'clock were expecting Nebraska boys. Naturally we had candy, cigarettes, etc., but we very willingly gave those things to the Kansas boys.

Smiles, tears and laughter followed. Appreciation showed

on over 300 faces. An officer told me it was the first time anyone had met their train and that North Platte had helped the
boys keep up their spirits.

*I say get back of our sons and other mothers' sons 100 per
cent. Let's do something and do it in a hurry! We can help
this way when we can't help any other way.*
—RAE WILSON

The men and women who helped with the Canteen—
and the men and women who, serving in the armed forces
of the United States, arrived at the Canteen—were in their
late teens, their twenties, their thirties and beyond in the
1940s.

Now it was sixty years later; now it was the twenty-first
century. Most, almost undoubtedly, were dead; those still
living were old men and women. I didn't know exactly
how I would find them, but coming to North Platte had
to be the only way to start. The people of this part of
Nebraska who had volunteered in the Canteen—it was
unlikely that they had all left. This had to be the part of the
country to find them.

And those who had passed through? The soldiers,
sailors and airmen who were on those trains?

I didn't know. But there might be records here—ways
to find the names of some of the military men who had
rolled across the middle of the United States, and who had

found themselves stopping for ten minutes in North Platte. From those records, perhaps I could start to locate the soldiers who were still living.

And then there was the Canteen itself.

It was gone. I arrived in town, and went to the place where the train depot had stood, and there was nothing there.

It had been torn down in the winter of 1973. Two years before that—on a day in May in 1971—the last passenger train had arrived at, and then departed, the Union Pacific depot in North Platte. The old passenger railroads were no more; the newly devised Amtrak had replaced them, and Amtrak's management had decided that its sole route across Nebraska would not pass through North Platte.

So the Union Pacific had demolished its station on Front Street. I went down to where it had been; all that remained was a plaque and a flagpole.

This was going to be like looking for a ghost.

Four

My first morning in town I awakened with the sun and decided to walk with no set destination.

I knew that I should head in the opposite direction of the interstate; I-80 was designed and built with the express purpose of moving people past towns like North Platte as quickly as they could drive. The interstates were meant to be the national bypass—the way to avoid taking things slow, to avoid looking the old America in the eye. They were a federally funded averted glance.

So with my back to I-80, I followed a path that led to a bridge that spanned the South Platte River, then turned left on a street called Leota. After a few blocks I passed the Great Plains Regional Medical Center, a modestly sized but relatively new hospital complex. Centennial Park, with baseball diamonds and skateboard ramps, abutted the medical center; I passed the Adams Middle School and a church with an empty parking lot, and soon found myself at the end of Leota, with a new housing development going up within eyeshot.

What I saw in more than a few front yards were satellite dishes—pointed toward the heavens, poised to bring the outside world to North Platte. From the north came the faint sound of a train whistle, but I knew it had to be a freight train—trains that carry people did not come here anymore, did not even run through here.

After about forty-five minutes of walking I turned around. The temperature again was already close to one hundred degrees, and few people were out on the street. I did hear a single voice, that of a child.

"Look what I'm making," the high-pitched words announced.

He wasn't talking to me; he must have been three or four, and he was out by the open-doored garage of a house on Leota. He was speaking to a woman who appeared to be his grandmother; he was showing her something that he was doing with wooden blocks.

I stopped at the sound, and looked over at them. The grandmother saw me, and looked back with a question mark behind her eyes. I didn't blame her. I was a stranger in town, standing outside her home.

I walked on, hoping my presence had not upset them.

There was a time when no one was considered a stranger.

"I was a high school girl when the war started," said Rosalie Lippincott, seventy-four. "I was a cheerleader and a member of the Pep Club. As a young farm girl I grew up happy, healthy and safe, preparing for the best of the future with excitement and enthusiasm."

She was one of the first of the Canteen volunteers I was able to locate. Today she lives not in North Platte, but in the Nebraska town of Central City; she told me that when she was a girl, the nearest town to her father's farm was Shelton, population eight hundred.

"On Saturday nights, my folks would take the cream and eggs from the farm into Shelton," she said. "For trade. That's what my dad would do—trade the cream and eggs for groceries. For sugar, and the staples.

"My dad would get his hair cut in town on Saturday night, too, but the real reason was to bring the eggs in. We might bring twenty-four dozen. The cream we brought to town was in five-gallon cans. They would be heavy, but my dad was strong.

"Saturday night was the social time of the week on Main Street in Shelton. While the men were getting their haircuts, we kids would go to the picture show at the Roxie Theater. It was usually a cowboy movie, and when it was over, if we still had ten cents left we would go to Rex Honnold's Rexall Drugstore for an ice-cream cone or a candy bar.

"For a country girl, the drugstore was kind of an exciting place on a Saturday night. The marble-topped fountain, with spigots that poured out mysterious, bubbly drinks . . . the little tables with the drugstore chairs . . .

"You might pick up a nice little gift for your mom or your sister in the drugstore, if you could find something for a reasonable price. But what we always kept in mind, all evening long, was that our parents had told us: 'Now, when the streamliner goes through town, we're all going home.'"

The streamliner, she said, was the *Challenger*, en route from Denver to Chicago. "It would come through Shelton around ten P.M.," she said. "The tracks were on the south side of Main Street. And Main Street was really only one block long. The train didn't stop. We knew it would never stop in Shelton.

"You could hear it from a distance. It came from the west, and it came through fast. The street would be lined with cars—people sitting in the cars, usually the women visiting with each other. And that train would come through—it looked like a streak. Sixty or seventy miles an hour.

"Oh, how many Saturday nights when I watched that train speed through town, and I wished I could trade places with someone on the train. That's all that I wanted."

But the train was gone within seconds, and her family was waiting for her at their car, for the ride back to the farm.

Which is why—when the North Platte Canteen started, and a group who called themselves the Shelton Ladies volunteered to do regular duty there—she almost leapt to offer her help.

"We got to ride on the train," she said. "It wasn't the *Challenger,* but it was a train. I would sleep overnight at a friend's house in Shelton, and we would run the eight blocks to the station. The train left Shelton at three twenty-five A.M.

"We would take hard-boiled eggs with us. In the Canteen, we would peel them and make egg-salad sandwiches. So many sandwiches—twenty bushel baskets lined with clean towels, with all the sandwiches waiting for the soldiers. It was really quite overwhelming—we would put magazines out for the soldiers. *Life, Look, Liberty, The Saturday Evening Post, Reader's Digest,* comics, movie magazines. We put Bibles out, too. Everything was free for the soldiers to take with them.

"The rest of the Shelton Ladies were . . . well, they were ladies. Women. But my friend and I were teenagers—and boy, was that a thrill, to be in the middle of all those guys in uniform!

"We would hear the call of 'Troop train coming in!'

and we would hurry to make sure that all of the fried chicken, pickles, fresh fruit, sandwiches, hot coffee, cookies, milk and chewing gum were out on the U-shaped table. And then the soldiers would come running in.

"Oh! The different accents, the different colors of skin . . . the men in khaki, the boys in Navy blue, the Air Corpsmen, the Marines, the dogfaces, the ninety-day wonders, and those with ribboned chests. And all of this for just ten minutes at a time! I think it was the beginning of opening my eyes to a bigger world.

"They would hesitate, and say, 'How much do I owe you?' When we would say, 'You don't owe us anything,' they could hardly believe it.

"There were no paper products back then, the way we know them now—they'd take their cups of coffee onto the troop train, and at the next stop someone would gather up the cups and put them on the next train back. It was done on the honor system, and we would always get the cups back.

"So you had all these dirty coffee cups that had to be cleaned and washed. I don't remember there being any automatic dishwasher. We did them by hand. We would fall asleep on the way home that night."

She told me that she did not, at the time, give much thought to where the soldiers who had eaten her egg-salad sandwiches were going. But something that she saw one afternoon on her dad's farm made her understand.

"There was a B-17 bomber that crashed near my home," she said. "It was on a training flight. It just started spiraling—you could see it spiral, spiral, spiral. . . .

"I was out in the fields with my dad. We were irrigating—I was sort of my dad's hired man. It was a hot August day. We saw the flames, and we went to the house and called the telephone operator. It was a hand-crank phone—she said she had just heard from someone else. The plane had gone down in a cornfield. It was a four-engine bomber, and ten men died.

"When I think about those years now, I guess I always knew darn well that a lot of those soldiers in the Canteen were never coming back. But I was an optimistic girl, and I tried not to let it enter my mind. They were handsome young men in the Canteen, people I admired, and I wasn't going to think about them not coming home.

"That B-17 spiraling, though . . . right on the farm, right on an August afternoon . . .

"There's a big world out there. It's not all just Main Street in Shelton on a Saturday night. I think that's what all of us learned."

That first full day in North Platte, I stopped in at the Lincoln County Historical Society and began to go through some records—old guest books that had been saved from the Canteen, in which soldiers had signed in and had listed their hometowns; letters of thanks that over the years had been sent to North Platte (often just to the town itself—no name or street address; the writers were expressing their gratitude not to an individual, but to a city); documents from the Union Pacific Railroad.

It became evident to me right away that it would be possible to find some of the men who had been on those trains. It would take some doing—but some of them still had to be out there.

Rosalie Lippincott and her friends who had made the egg-salad sandwiches and had neatly arranged the *Life*s and *Liberty*s on the card tables . . . they were one side of this.

But if ever there was a two-sided tale, this was it. Those men who had spent their ten minutes in North Platte . . .

Who were they? What must they have thought?

"On the sixth of June in 1944, I was still in high school," said Russ Fay, seventy-five years old, when I located him at his home in Greendale, Wisconsin. "Pulaski High School, in Milwaukee. But boy, the Army grabbed me the next day.

"I was inducted at Fort Sheridan, north of Chicago. And then right onto a troop train, on my way to basic training at Camp Roberts, in California.

"Two men slept in the lower berth, and the guy in the upper berth got to sleep by himself. We rotated every night. We were typical eighteen-year-olds . . . we were in the Army now, and that was that. I don't remember a whole lot of moaning or groaning.

"On the train ride across the country, I had absolutely no idea what was going to happen to me. Like most of the guys, I was hoping that I didn't end up in the infantry. We had all heard that in the infantry you had the most chance of getting killed.

"There was a mess car on the train. Pretty soggy food, the same food every meal. We just sort of accepted that it would be that way the whole way.

"And then we rolled into North Platte.

"I had heard of North Platte in school, in history class, because of the Overland Trail. The pioneers. But who ever thinks they're going to be there?

"The train stops in North Platte . . . and we see these women carrying baskets toward our car. It's the daytime, it's hotter than blazes, and we can see that there's sandwiches and things in the baskets.

"We don't understand it. The women get onto the train—for some reason we weren't allowed off—and they're offering us the sandwiches, and these little glass bottles of cold milk.

"We're seeing more people out on the platform—from what I recall, there was a wide area out there. We're asking them why they're doing this, and they're telling us that they meet every train, every day, every night . . . and I remember how much I appreciated it, and especially I remember that the sandwiches were so good.

"I found out later it was pheasant—these delicious pheasant sandwiches, with mayonnaise. I can still taste it. Can you imagine that? Ladies are coming onto your train, and they're giving you pheasant sandwiches?

"The fact that they were women . . . I mean, in the 1940s, you're eighteen, but you're still a kid, at least before you get to the war. The women are on the train—and you might whistle, but that's all. You didn't really flirt.

"We were there for such a short time. The rest of the way, we kept thinking that maybe there would be other places like that. We wanted it to happen again. But it never did—Utah, Nevada, it got pretty desolate, and we'd stop to take on water and coal, but no one ever met us. We never ran into anything like that before, or after.

"I ended up in France and Germany, in an infantry division, field artillery. And all the way over there in Europe, across the ocean, you would hear about North Platte. Guys would mention it, guys you hadn't met until you got to Europe. We were scraping the bottom of the barrel for food, eating field rations, and someone would say: 'I wish we had some of those sandwiches like they gave us in North Platte.'

"You have to understand—the memory of food was such a powerful thing over there. I would see fights break out over it. Someone would get carried away describing a wonderful meal they had before they went in the service. Sort of needling the other guys. It was like a form of mental torture. A guy would say 'Shut up,' and the guy would keep twisting the knife about the great meal, until the guy he was needling couldn't stand it any longer.

"I lost sixty pounds overseas. I weighed one hundred eighty going in, and I came out weighing one hundred fourteen. My mother didn't even recognize me at first. You talk about the Jenny Craig diet—there was nothing like the World War II diet.

"I would say that a majority of the men on the battlefields knew exactly what North Platte was, and what it meant. They would talk about it like it was a dream. Out of nowhere: 'How'd you like to have some of that food from the North Platte Canteen right about now?' "

He came home after the war, he said, took a month off to do nothing, and then went to work at the Allen-Bradley electrical equipment factory in Wisconsin, where he remained for the next thirty-eight years.

I asked him if there was anything he would like me to tell the residents of North Platte.

"Just that I don't know how those people kept it up for all those years," he said. "Especially with all the shortages during the war . . . how did they do it?

"It's funny what you remember. When those ladies came onto the train, I remember that there was a real big napkin in the bottom of the basket, and the sandwiches were laid out on the napkin.

"Just tell them that I still thank them from the bottom of my heart. And that if they ever ask themselves whether what they did really mattered, that the answer, to put it bluntly, is: Hell, yes."

That night in my hotel room, the weathercaster's voice came out of the television set, giving the local forecast for the next day:

"Hot, hot and more hot."

But in the midst of a summer of soaring heat, my thoughts kept going back to that Christmas week in 1941, to the freezing Nebraska nights when the Canteen began. For the few people there at the very start, before anyone could know how long the war would last, and thus how long their commitment would be required . . .

Before it was history—when it was just a silent impulse out on the plains, in a tiny place cut off by time and distance from the rest of the country—who from the town walked into that railroad depot, and what, in their own hearts, brought them there?

Five

"*I had read* about it in the paper—about Rae Wilson needing helpers for this project she had in mind—and so I went down there on Christmas night, the first night the Canteen opened. I was twenty-three and a single woman, and I was lonely that Christmas. So I went to the railroad station. There were only five of us there that first night."

LaVon Fairley Kemper, eighty-three, who now lives in Littleton, Colorado, was telling me the story of that night in 1941 when the Canteen began in earnest.

"I was teaching school in Lodgepole, Nebraska, about seventy-five miles west of North Platte," she said. "I had come to North Platte to spend Christmas with my dad.

He was single, and he was living in a rooming house. We had no Christmas tree—we really had nothing. He had a room, and I slept on a couch in the living room of the rooming house.

"We had finished our Christmas dinner, and the paper had said that Rae Wilson was looking for volunteers, so I told my father I would be back. It must have been eight-thirty or nine o'clock when I came down to the depot. I was feeling pretty lonesome that night.

"At the depot, there were seven or eight bushel baskets that the ladies had filled with apples and candy. They had stored the baskets in the lobby of the Cody Hotel, because there was no one to guard the apples on the train platform, and as cold as it was, the apples probably would have frozen. By the time the train came in, there wasn't a soul at the depot except for us.

"It didn't come in until about eleven o'clock that night. The news about troop train movement was very hush-hush. Later, after the Canteen was up and running, we would get the word from railroad people—the code we used was 'The coffee pot is on.' Meaning a troop train was on the way.

"That first night, the soldiers on the train were so amazed—way out there in the boonies at eleven o'clock on Christmas night . . . they were quiet as they looked out of the train at us. We carried the bushel baskets out to the train, gave the men the apples and the candy, wished them Merry Christmas, and the train left.

"I think after that we all told each other goodnight, and we went our separate ways. Rae Wilson told us to come back the next morning, which was fine with me, because my dad worked days. He was a machinist."

She helped out at the Canteen all that Christmas vacation, she said, and then went back to Lodgepole to teach. "It was a town of only about two thousand people. I taught history and English and Latin at the high school, and I coached the high school plays and was the adviser for the Pep Club. Life was so sort of empty in Lodgepole, with no men in town because of the war, that I just took on everything I could do at the school.

"They had an assembly that first year of the war, and I was looking out at the boys in the auditorium and thinking, these boys are going to end up going to war, and some of them are not going to come back. I tried to keep the tears from flowing down my cheeks.

"I would go to North Platte one weekend a month, to stay with my dad and help out at the Canteen. It was very, very quiet in Lodgepole, with not very much for a young woman my age to do. But I wanted to show an interest in the town that was providing my living, and not have them think that I was just a carpetbag teacher who would leave every weekend. I thought I owed it to them, to the small town, to be there, to be in the church on Sunday, to get to know the parents of the boys and girls I taught.

"I guess I looked forward a lot to my one weekend a month visiting my dad and working at the Canteen. For that one weekend a month I would be around the fellows, and I felt I was doing something for the war. It wasn't so lonely."

One thing she will never forget, she said, was what happened to a woman who volunteered with her at the depot—a woman by the name of Elaine Wright.

"Her husband was a railroad man. He was one of the ones who knew when the troop trains were coming, and who would tip us off. She had a son, who was off in the Navy.

"Most of the older women who worked in the Canteen had sons in the war. It was like a healing thing for them, to work there—their homes felt hollow with their sons away, and I think they sort of built their world around the Canteen.

"I think Elaine Wright was in the Canteen when she got the word that her son had been killed in action. I wasn't there that day, so I don't know for certain that she found out inside the Canteen. But I believe that was so.

"What I do know is that after being away for several days because of her son's death, when she came back to the Canteen you could hear a pin drop when she walked in. There was silence, and a lot of hugs. And then she said: 'I can't help my son, but I can help someone else's son.' And she was there day after day.

"I didn't see any change in her after her son's death, except that she was probably even more caring to the boys who came in from the trains. I don't think she ever told any of those boys that she had lost a son."

LaVon Fairley Kemper told me that after the school year was over in Lodgepole, she moved to North Platte, got a job teaching at Roosevelt Elementary School, and persuaded her father to move out of the rooming house where they had spent that first Canteen Christmas, and to rent with her an apartment that they could share: "It was a nice place." And she continued to volunteer at the Canteen.

The Christmas alone with her dad, the sharing of the apartment . . . I asked her if her mother had died, and if her dad had been a widower.

She paused for just a second.

"My mother was a come-and-go mother," she said. "She changed men frequently. My father was a quiet, gentle man. I don't know what my mother being that way did to him inside.

"My mother . . . I think she would have made a fun sister. But not a mother. When you go off and leave your daughter . . . she shipped me off to her sister when I was in the second grade. Then again when I was in the sixth grade. She really didn't care. Her fly-by-night life was more important to her than I was. She would just announce that she was leaving, and that she had made arrangements for me.

"My mother died when she was sixty-two, which was probably just as well. She would have made an awful old woman. My dad never remarried. I never knew him to go out with anyone. I just admired him so much and loved him so much. He was the one thing that stabilized my life. I depended on him completely.

"He lived to be eighty-five years old. He died in 1977. He stayed in North Platte, after I was married and I was living in Colorado. I made sixty trips to see him in the last five years of his life. An eight-hour trip each way. I would catch up with his business and do his mending. And I would sit with him and tell him what a great father he had been to me.

"Those trips were my payment to him. For being the man he was, and the father he was."

She made those sixty trips from Colorado to North Platte and back by bus, she said.

"The passenger trains were gone by then," she said. "There were no trains that would take a person to North Platte."

I noticed it right away:

People waved from their cars.

I was making the ninety-minute walk my daily routine—from the Quality Inn across the South Platte River, over to Leota, up past the hospital and the church, around the bend to E Street and then turn around and head back—and the waves from the front seats began to seem automatic.

You wouldn't necessarily think that would be the case—certainly I had understood the wariness from the grandmother that first day in town, when I had paused by the house where her grandson was playing with blocks. These weren't heavily traveled streets, I was someone who hadn't been around before. . . .

Maybe it was the safety that people feel inside their cars. They were moving faster than I was—maybe there was no gamble in a quick wave to a stranger, or a sociable beep of the horn, which I was hearing all the time, too.

Still—it's not something you experience much in a big city. Especially if you cross in front of a car whose driver thinks he or she has the right of way, the only wave you're likely to get consists of a single raised finger. And the horn honks directed at you are not often of the amiable variety.

Here, though, it had a "Good morning" quality to it. I couldn't help thinking about whether any of this—the small-town instinct by motorists to greet someone they've never met—might be a remnant from an earlier time in the town's history. A time when the greeting of strangers was the town's stock-in-trade.

Because I was finding more and more of those long-ago strangers—men who had been here many years before I, men who had come briefly to town in a previous century—who remembered the greetings quite well.

"I didn't know North Platte myself," said Jack Manion, seventy-eight, who now lives in Florida, north of Tampa. "But it so happened I had a high school sweetheart who was living there."

Manion had been in basic training in California in 1943, preparing to serve in the Army Signal Corps. A troop train was scheduled to take him and his fellow soldiers all the way across the country, to New Jersey. He had sent a telegram to his girlfriend—he had not seen her for more than a year—saying that he would probably be stopping in her town for ten minutes or so. He was not permitted to say what train he would be on, but he indicated what day he would likely stop in North Platte.

"On the train, I got slicked up as best I could," he said. "I told the other guys: 'Put your shirts back on, get decent.'

"But I couldn't get the guys too interested. There's a lot of lethargy that builds up when you ride on a troop train. A lot of the guys couldn't believe there was anything special for them in a little cow town.

"I knew I wanted to look nice for my girl, but that wasn't an easy thing to do. You packed in barracks bags in those days, and your clothes got pretty rumpled. You had two bags—you put your laundry in one bag one week and in the other one the next week. You would try to have your clothes not be wrinkled and your shoes shined—we were as serious about that as twenty-year-olds can be.

"I was trying to keep myself cool on that hot train. I was doing my best to try not to perspire through my clothes.

"We pulled into North Platte, and the guys couldn't believe all the people waiting for them on the platform. I got off with the rest of them, looking all around for my girlfriend. . . .

"And there she was. She was out there kind of looking for me. I was one of many guys coming off that train wearing the same khaki uniforms. I can't even remember if she had seen me in a uniform before. But we found each other. She had driven to the depot, and I guess she had waited for a number of trains, thinking I might be on one of them.

"Right next to the curb was her Buick. Black. Parked parallel, right next to the depot.

"We got into the backseat of her car. We were . . . well, we weren't, you know—but we were having a private session, getting to know each other again. People were looking at us, but we didn't care. I knew I only had a few minutes, and I was afraid I'd miss the train. If people walked by and stared at us, I wasn't going to waste time thinking about that.

"Some of the guys came by the car—they were saying 'Way to go, Jack!' It was probably an awkward situation for her. We had kind of drifted apart. We hadn't separated, but we hadn't seen each other in a long time, either.

"I believe it was a train whistle that let me know it was time to go. I looked out of her car, and the troops were dispersing from the depot, and I sensed that our time was up.

"I dragged her with me through the Canteen—it was thinning out quite a bit, the guys were almost all back on the troop train. She was only nineteen or twenty—she was probably a little embarrassed. I put my arms around her as we stood on the platform, and we were kissing goodbye. All the guys were sticking their heads out the train window—some of them were polite enough to ignore us, but most weren't. Like I say—this had to be embarrassing for her, but she hung in there."

He ended up serving in the Pacific, he said, on Saipan; after the war he was hired as an electrical engineer by General Electric in Syracuse, New York, where he worked on, among other products, color television sets. That life was what the train had ultimately been taking him toward.

And the girl in the black Buick at the North Platte train station? Did he come home from the war to marry her?

"Alas, it turned out someone else did," Manion said. "They say you outgrow each other. We wrote to each other for a while, but pretty soon that stopped. We weren't mad at each other. We just drifted apart, as people do." He doesn't know what happened to her. He met someone else—and they have been married for fifty-three years.

When the troop train pulled out of the North Platte station that day in 1943, he said, "It was kind of lonely. A little sad."

But he knew that, if he was fortunate, someone else awaited him down the tracks, toward the east.

His father was a civil engineer: "His job was such that we moved around a lot." And at this point, his dad had a job near the Nebraska town of Kearney. Manion had let his parents know he most likely would be traveling past Kearney on this day.

"We weren't going to stop in Kearney," he said. "But between the cars of the train, there was this little platform. They let me stand on it as we approached Kearney.

"My parents were waiting there, next to the tracks. As the train rolled through Kearney, it slowed down a little bit. I was able to reach my hand out and shake my father's hand. Just for that one little moment.

"The conductor of the train was nervous about it, but he allowed me to do it. It was: Here I come, here I go. That brief. I think my father kind of had tears in his eyes as our hands touched.

"But he knew that it was a troop train. He knew that it couldn't stop."

Six

On the weekend after I arrived in North Platte, the *Telegraph* devoted a full page to coverage of young men and women in the area who had decided to become husbands and wives.

Each engagement story was accompanied by a photograph of the couple; the men and women were dressed casually, for the most part. Open shirts (even T-shirts) for the males, everyday dresses or blouses for the females.

About Angela Faye Abbott of Kearney, and her fiancé, Shawn Henry Warner, also of Kearney, the newspaper reported: "Abbott is a 1996 graduate of Hartington High School and a December 2000 graduate of the Nebraska

College of Technical Agriculture in Curtis, with a degree in horticulture systems with landscaping, greenhouse and turfgrass options. . . . The prospective groom is a 1995 graduate of Ord High School and a December 2000 graduate of NCTA with a degree in natural resources. He is employed by Buffalo County Surveying in Kearney."

Love finds you. Out here on the plains, or in the commercial canyons of New York or Chicago, love finds you. Michelle Blaesi of North Platte, who, according to the *Telegraph,* attended the Mid-Plains Community College dental assisting program and who was working as an assistant manager at the Subway sandwich shop, would be marrying Joshua Nelson, also of North Platte, who had attended Nashville, Tennessee, Diesel College, and who was working at Guynan's Machine and Steel.

Love finds you anywhere. One of the engagement announcements—because of the small town where the prospective bride and groom lived—caught my eye especially. Renee Munson and Luke Connell, both of Tryon, would be getting married; she, according to the *Telegraph,* was employed at Wal-Mart, he worked "as a ranch hand for Ron Lage."

Tryon was a tiny place with which I had become familiar because of one woman who, during the 1940s, had grown up there. Her name was Ethel Butolph, and now she was seventy-nine. Love will find you. Yes it will.

"The Canteen?" Mrs. Butolph said to me. "I never went there. But I got my husband there."

Tryon, she said, was about as small as a town could be: "Seventy or eighty people. In McPherson County, about thirty-five miles north of North Platte. There was a post office and a little grocery store and a high school, but that was about it."

In her family, she said, there were ten children—nine girls and a boy. She never went to high school. "I guess I thought I was smart enough. Not every girl went to high school back then."

Women from Tryon volunteered a regular shift at the North Platte Canteen, and they came up with a tradition. The women would make popcorn balls to give to the soldiers, and in each popcorn ball they would put the name and address of a girl who attended Tryon's high school. The idea was to give the soldiers someone to write to.

"Now, today, a girl would probably be scared to do something like that—write to someone she has never met," Mrs. Butolph said. "Then, it was a friendly thing to do."

But the popcorn balls at the Canteen didn't apply to her—at least they weren't supposed to. Because she wasn't in school, she wasn't thought of when the Tryon Ladies put the names in the balls. "I was just staying home most nights, doing some baby-sitting for people, and I had a job in the post office over in Ringold. Ringold was even smaller than Tryon. I got the job because all the boys were in the service. The Ringold postmaster went off to war, so I took his place."

Her younger sister Vera, though, had her name put into a popcorn ball, because Vera was in school. "A soldier named Woodrow Butrick got the popcorn ball with Vera's name in it when he passed through North Platte," Mrs. Butolph said. "He wrote to her. The first letter, I believe, was from Ogden, Utah. These boys were lonesome. They would do anything to get some mail."

Vera and Woodrow began to correspond regularly, although they had never met or spoken. And after a while, an Army buddy of Woodrow's said to him: "Woody, you don't know if she has a sister, do you?" The Army buddy was a young man by the name of Virgil Butolph.

"So he began to write to me. He was four years older than I was. Vera and I would just write friendly letters back to the two of them. Mom would read the letters before we mailed them, and she also read the letters that Virgil and Woody sent to us.

"It went on for three years. It got to the point where my sister and I were writing to them every day, and they were writing to us just about every day, too. Woody went to Germany, and Virgil went to Alaska, and we kept writing.

"We sent them pictures, of course, and they said they liked the way we looked. The way we felt about each other grew by letters. I think you get to know a person better through a letter than you do by seeing them. When you see someone in person, you're putting on such an air. In a letter you tell things the way they really are.

"We corresponded from 1941 to 1944. He sent me a ring at one point, and I said in a letter, 'I'll wear it, but this doesn't mean we're engaged.' He wrote back and said, 'OK.' But I think we realized that maybe this would be it—maybe it would happen.

"I received a telegram from him in 1944. It said: 'Coming home.' He came to Nebraska on a furlough. He got off the train at Kearney and he borrowed a car and he drove to Tryon. I thought he was very cute, when I met him.

"For our first date we went to a rodeo over at Stapleton. We went with four of my sisters. My parents didn't seem to mind him.

"We were married two weeks later, while he was still on his furlough. We got married in the North Platte Methodist Church. Then he had to go back to his base in Alaska.

"When the war was over he came back to Nebraska and got a job in North Platte with the railroad. For the next thirty-two years we lived very happily together. We had five children.

"The day he died, we had gone to look at the cranes. They come here every spring and fall. This was on the cranes' spring stop here. We had driven out to see the cranes, out in the country, near Hershey. There must have been about a thousand of them, out in the cornfield, doing their dances, jumping up and down. Virgil and I always loved to drive out and watch them.

"We were driving home. I remember, we were on a street with no traffic. He hadn't been a bit ill. It was about two in the afternoon. He slumped over the wheel; it was a heart attack. By the time the emergency workers arrived, he was gone. He was fifty-nine."

She was silent for a second, and then said: "Don't make it sound too flowery. It's been a good life. I have no complaints."

She wishes she could have seen the Canteen just once, she said, because of how it had brought her husband to her. "During the war, you couldn't get gas for a trip like that," she said. "Those thirty-some miles from Tryon to North Platte, and then back again . . . you didn't make a trip that long unless you really had a reason you had to do it.

"But I wish they hadn't torn the depot down. I don't know what I would have come to, if it hadn't been for that place. The Canteen changed my whole life—and I never even set foot inside it."

Love will find you—apparently it has always been true. Love will find you, when it is least expected.

One afternoon when I was in North Platte, I spent several hours going through old steel cabinets where letters, documents and news clippings from the Canteen years were stored. I found this—sent in April of 1944 to Mrs. S. C. Rabb, of 514 North Maple in North Platte.

The letter had been written by the mother of a young airman who had passed through the Canteen, and who had been given a cake that Mrs. Rabb had baked.

The boy's mother, Ethel Koncilek, of West Sayville, Long Island, New York, had written to Mrs. Rabb, whom she had never met:

> At a stop in Nebraska several days ago, a dusty, hungry, travel-weary teen-age air cadet stepped off the train to relax. He had started in Nevada for home, which he was to visit for the first time in eleven months. He was still a thousand miles away. You know what happened at the brief stop in Nebraska.
>
> Please accept this letter as a token of our appreciation for your kindness to our boy. May God bless you and your loved ones. This world may still be a better place in which to live as long as charity such as yours remains in the hearts of men.

And then there were the young soldiers who were in search of . . .

Well, if not love, then the next best thing.

"I was a farm boy from Wisconsin," said Don Griffith,

seventy-eight, when I located him in Indian Harbour Beach, Florida. He laughed. "I was getting ready to join the Sixty-sixth Infantry, and when we were training, we were out to meet girls and get a free dinner. In that order."

He said that he had enlisted when he was nineteen, and was receiving Army specialized training in engineering; the course was being given in Golden, Colorado.

"We lived in fraternity houses on a college campus, and we went to classes at the college," he said. "We were very military—when the professors walked in, we would jump up and stand at attention. I think we scared them.

"But what we really wanted to do was meet some girls. For some reason, we told ourselves that the best girls to meet were over the Colorado line, in the rural part of Nebraska. Don't ask me why we were convinced of that.

"We would take the Galloping Goose streetcar line from Golden to Denver. Then we'd hitchhike to Nebraska, and we'd head for church. I'm not kidding you. That was one way to meet girls—find them at church.

"Now, these girls in church were pretty well chaperoned. And . . .

"Aw, all right—I'll tell you why we thought we had a better chance in Nebraska. We wanted to get far enough away from where we were training that the girls we met would not be so used to seeing guys in uniform. We would stand out.

"So we'd meet the girls in church, and they would

invite us back to their farm or ranch to have dinner with them and their parents. . . . It was nice. We were away from home. We were meeting girls. Yes, we had to hitch-hike to Nebraska and go to church to do it. . . ."

He laughed again, and then turned serious.

"We may have been young," he said, "but we weren't too young to appreciate what happened to us at that North Platte train station." He had passed through North Platte after his training in Colorado was finished.

He said that although the food and the coffee he was given at the Canteen were tasty and filling, "That wasn't the biggest thing about North Platte. The biggest thing was how those people made you feel really appreciated. Those happy smiles that you saw. They were just being so nice.

"I know it sounds like a simple thing. But I was heading for an infantry division when I went through North Platte, and I didn't know exactly where I would end up [he would end up in France]. And I never forgot those smiles.

"The men in North Platte were mostly gone, like they were in every town. That one little town could never have supported the Canteen all by itself. It had help from all the farm communities, all around the area. They all came to North Platte to make sure no train ever went unmet.

"You don't forget something like that, when you're overseas. There was no place else I ever knew of, or ever heard about, that went to that great effort. A lot of people might be *willing* to do it. Or at least they might say they would be willing. But in North Platte, they *did it*."

Mr. Griffith said that he finds himself thinking about his brief stop at the Canteen more than he would have imagined. "My generation is disappearing," he said. "I doubt that our children or grandchildren will know that North Platte ever existed. That was the *other* side of the war—the one that doesn't get mentioned in the history books. What the people at home did. How supportive the civilian population was.

"We were masses of soldiers. There were almost more of us than there were civilians—at least I'll bet you there were more of us than there were men our age who didn't go.

"The men of our age group who didn't go probably never got over it. They missed the experience we had. All of us came home—and maybe you had a cousin who had asthma, and he didn't go into the service. We'd come home and have our stories and our drinks, and the ones who didn't go . . . that guy would be left out.

"It wasn't only the battles they missed out on. They missed out on something like North Platte—they missed out on knowing how good people really can be, how considerate. It wasn't easy, what the people at that train station did for us—as I remember, we came through in the middle of the night. The troop trains didn't time it to make it easy for the people who worked at the Canteen. The trains came when the trains came.

"And the people were *there*. You should have seen it. You have no idea what that meant to us. The middle of the night. And they were *there*."

Mr. Griffith was accurate about the people in surrounding towns and counties making sure that the Canteen never went unstaffed. I found a list still on file in North Platte— a list of the communities whose citizens regularly traveled to the Canteen to help.

Absolutely remarkable. More than 125 communities, not just from Nebraska but from Colorado, too. An honor roll of towns: Anselmo, Berwyn, Bignell, Brandon, Dry Valley, Dix, Oconto, Lillian, Sarben, Roscoe, Lewellen, Tallin Table, Thune, Lemoyne, O'Neill, Verango. . . .

Some towns you can't even find on a map, sixty years later. Towns so small they have disappeared. But during a time of precious gasoline, and no interstates to make the journey smooth, the people from those towns got to North Platte. They saw to it that when the trains pulled in, someone was there.

Holbrook, McGrew, Elsie, Brule, Bucktail, Farnam,

Flats, Arthur, Birdwood, Sunol, Wallace, Westerville, Eddyville, Elm Creek . . .

They dropped everything they were doing, and when the young men looked out the windows of the trains, the smiles were waiting for them.

Love will find you. It will.

Seven

There was a photo around town—an old black-and-white picture taken at the Canteen—that I had seen more than once. For older residents of North Platte, this particular photograph seemed to sum up much of the spirit of the town during the war years.

In the photo, a young sailor—you can see him only from the back—is seated at a piano in the Canteen, apparently playing away. Surrounding him are other military men and women, all in uniform, some sipping coffee. And above the piano—as if staring down at the sailor making the music—is a caricature of Adolf Hitler.

Playing and singing in the face of war and death—what

a photo. So I was surprised, one day in town, to meet the sailor in the picture, now grown older.

His name was Lloyd Synovec, and he was seventy-three. He told me he had not always lived in North Platte—he hadn't when the war had begun.

"I was seventeen at the tail end of the war," he said. "I grew up in eastern Nebraska—a town called Pierce, about fourteen hundred people, near Norfolk. I came from a Navy family, even though we were midwestern. I would have liked to have gone into the service even sooner, but my mother . . .

"Well, you know how mothers are. She said, 'Why are you in a hurry? You can wait a while.' But I was seventeen when I went in, in forty-five.

"I was in boot camp in San Diego, and I was given the opportunity to spend a little time back at home, so I hitch-hiked east. Me and another kid from the base in San Diego. We got some rides up the coast, and then, right outside of Reno, we got a ride the rest of the way from a wealthy guy who was there for a divorce. He was from Bayonne, New Jersey. Believe it or not, people at that time picked you up if you were hitchhiking. They'd go out of their way to give a guy a ride.

"He had a big Buick. For a seventeen-year-old kid from Nebraska, that was like a limousine. We even helped him drive. He paid for the food. He had been out there getting rid of a wife."

After Synovec's visit to his family in the eastern part of Nebraska, he started back to the base in California, riding on a passenger train. "I think it was in July or August," he said. "The war was pretty close to over, but no one knew it at the time. We didn't know the A-bomb was going to come.

"I knew about North Platte, and the Canteen. The fellow who had given us the ride home had driven through the town, on Highway 30, but we hadn't stopped and I hadn't seen the train station. But I knew about it, because my mom had sent a little money or something down there. Women from hundreds of miles around were baking cakes and sending them to the North Platte Canteen.

"The train heading west pulled into the North Platte station, and the conductor yelled that we had about fifteen minutes. There were other military people on the train, and we jumped off and went in. My first impression was: 'Holy cow!'

"You just didn't see things like that. The place was absolutely packed—in addition to the guys from our passenger train, there was a troop train stopped at the station, too, so the place was full of soldiers—more soldiers than sailors. I was kind of flabbergasted. I grabbed a sandwich and a glass of milk, and just looked around.

"There was a piano over in the corner. I had played a little bit in my high school's orchestra—I had studied music for six or seven years because my mom made me take lessons. I kind of quit for a while until I heard that girls liked piano players. Although, at seventeen, I don't think I would have known what to do with a girl.

"But at the Canteen I saw that there was no one sitting at the piano, and it looked like a pretty good one, so I took my sandwich over there and I sat down. I didn't expect to be the entertainment, but everyone started gathering around.

"I knew all the tunes of the day—'Mairzy Doats' and 'Praise the Lord and Pass the Ammunition' and 'I'll Walk Alone' . . . the other soldiers and sailors began requesting songs, so I played them, and they sang. I remember one that they called out for. . . ."

Synovec began singing to me, all these years later: " 'I left my heart at the Stage Door Canteen. . . .' "

He said that as he heard the soldiers and sailors around him singing during that brief stopover in '45, "I was pleased. Those of us who play piano are always pleased to find one that plays well, and this one did. I suppose it had been played by hundreds of guys who came through the Canteen during the war, and on this day it was me.

"I didn't even know that anyone took a picture. I didn't see it until years later. I had my blues on—I look at that picture today, at me in my uniform, and I don't think I could get my leg in now where my stomach was then.

"I remember they kept yelling what songs they wanted me to play. 'Let's hear "In the Mood." ' I wasn't nervous about playing the songs—I was nervous about *time*. I knew I'd have to get on my train before very long.

"Everyone was having a good time—you get that many people together, singing, there's no chance to be blue, at least for those few minutes. I remember that when it was time for me to go, I got up and made a little speech thanking all the ladies who made the food for us.

"I was pleased as I got back on the train. I'd played for the people, I had a big old ham sandwich and a cold glass of milk in my stomach—what more could I have asked for?"

He said he was stationed at Pearl Harbor at the end of the war, and served there for a few years after that. "I came back to Nebraska, and the GI Bill was available to all of us, but I decided I was not collegiate material. Looking back, studying was not my bit.

"I got a job as a Linotype operator—this was back in the hot metal days. I worked on weeklies at first—there was one called the *Fremont Guide and Tribune*. I worked on some commercial printing jobs, and I got a job at the newspaper in North Platte, where I worked for twenty-two years. That's how North Platte became my home.

"I'm retired now, unless you count mowing my own front lawn for free. I'm glad that I wound up here. This town has never been a big town, but it seems to have a good heart.

"I never would have guessed it, the day the train stopped and I went into the Canteen—I never would have guessed that I would end up spending most of the rest of my life here.

"That stop that day was so brief. But it's funny—I can still hear the music I played that day. I can hear it right now."

There did not seem to be an overabundance of entertainment options in town as Saturday night approached; I browsed through the North Platte *Telegraph,* looking for somewhere to spend a few hours, and came up empty.

I didn't want to just spend the night in my room. Around the hotel since my arrival I had run into quite a few men and women who told me they were serving as coaches or chaperones at a big softball event that was being held in town: the Amateur Softball Association's Sixteen-and-Under Regional Softball Tournament, in which girls' teams from around the Midwest were competing in an effort to make it to the nationals. The men and women had suggested that I come out and watch.

So, around dinnertime, I did. The Dowhower Softball Complex was on the north side of town, past the Union Pacific tracks where the Canteen had used to be. The local favorites—the North Platte Sensations—were scheduled to play the St. Louis Lightning.

The heat wave had not broken, not even a little, and the temperature still hovered around triple digits. The dirt on the back-to-back diamonds was baked solid; the coaches were telling the girls to "drink a lot of water," as if such a suggestion was necessary. The girls on both teams looked as though they had been on a forced march through the desert, and the contest had not even begun yet.

"You're the one, one, one, you're the one," the girls chanted in the direction of one of their players, trying to encourage her toward a good performance in the game to come. The team members were cheerleaders and participants, all at once—"They've got different cheers for every one," the mother of an infielder told me.

The star of the Sensations, I could see, was a pitcher the girls called Pook. A tall young woman, she had fire in her eyes as she warmed up. I was told that her real name was Jessica German, and that she lived in Cozad, some fifty miles away. She could have pitched for the Cozad Classics, but the North Platte Sensations were a much better team, a regional powerhouse, so she had chosen to play for them.

The game began; the parents of the players stood along the baselines, or, so they could see the right diamond, sat backward on the top row of a set of bleachers built facing another field. Their legs dangled in the air. A few feet away, a "Rain Room"—a tent with mist-producing sprinklers rigged up—provided a cooldown area for younger boys and girls, including a few in diapers who ran through the falling water.

Next to the home team's bench, a handmade sponsor board was propped up, featuring advertisements for local businesses that supported the team. The ads—pieces of paper inserted into clear plastic sleeves on the board—promoted companies that could serve a North Platte citizen on various stops from birth to death: the Old MacDonald Day Care and Nursery School. The White-Musseman Hearing Aid Center. The Carpenter Memorial Chapel, Troy Tickle, director.

"All right, Pook!" the girls on the Sensations called out as the game began. I found Pook's father—his name was Britt German—sitting atop one of the backward-facing bleachers, watching his girl. He told me that he was an electrical lineman for the city of Cozad, and that Pook was what Jessica had been called when she was a baby; the name had stuck.

He had encouraged her to play for the Sensations, he said, because "she loves the game, and this is a better team for her than the one in Cozad." It was not the most convenient decision; he had to drive her all the way to North Platte and back for every practice and every game of the year—each time a hundred-mile round trip. "I bring her and wait with the wives," he said.

He said he had never figured out just how much driving—and how much devotion—this entailed, but when I asked him, he did the math in his head. "One hundred miles a trip, maybe one hundred times a year," he said, "and she's been playing on teams here for six years . . . what does that come out to? Sixty thousand miles?"

"Two down, now!" one of the mothers on the baselines called to the team. She glanced toward the scoreboard, which was dominated by a Union Pacific freight-line logo with the slogan WE CAN HANDLE IT. The girls on the field peered toward the St. Louis batter, Pook German cupped the ball in her hand, her father leaned forward, and I realized that these girls at play were only two or three years younger than many of the soldiers who had once passed through this town on their way to war.

Those two or three years of difference—from mid-teens to late teens—must, in the days of the Canteen, have seemed like centuries to boys just too young to go to fight for their country. At least that is the conclusion I drew after talking with one of those boys, who had helped his mother at the Canteen.

"I would guess I was fifteen," said Floyd O. Berke, now seventy-six, who grew up in Eustis, Nebraska, about sixty miles east of North Platte, and who still lives there. "I was a member of the Tri-County Luther League, which was the organization of young people from churches in our area.

"My mother had gone to the Canteen with the ladies' group from our church, and when we young people volunteered to help, we met up at the church and each of us brought a gallon or so of gas to help us get there and back. We went six or more to a car.

"In addition to the gas for the car, we brought cookies, cakes that had been decorated, fried chicken, coffee makings, and breads, meats, spreads and cheeses for sandwich making. When we got to the Canteen the sandwich makers took a slice of bread, applied the necessary fillings, added the second slice and stacked the sandwiches to the side. When the stack of sandwiches was about twelve inches high, one person had the job of slicing them in half. He had a long knife and *carefully* sliced through the stack while the rest of us stood alongside to catch everything that might be dropping."

This was all in preparation for when the moment would come: when young men just slightly older than the sandwich helpers would arrive on the tracks outside. "When we were alerted that a train was coming, we got into our positions so we were ready to serve."

Mr. Berke told me that "for a bunch of kids as young as we were, it was fun. We were pretty much in awe of those soldiers and sailors in their uniforms, so close to our own ages. I had a lot of respect for them, just looking at them in the uniforms. The young and old soldiers were kind of mixed together—although when I think back to it now, the young soldiers were seventeen or eighteen, and the old ones were twenty-three or twenty-four.

"My job was to serve them coffee. I didn't know if I was doing it wrong or not. I just assumed that you let them pick up the empty cup, and then you filled it up while they were holding it. I think some of the time I filled the cup half-full before they got to it, so I could speed things up.

"It looked to me like they were having fun. I know that sounds strange, in a time of war—maybe I thought they were having fun because I knew they were going places I wouldn't have minded seeing myself.

"For the most part they were all cordial and polite. I think the military had a way of preparing soldiers for active duty so that the soldiers' minds could be set at ease a little bit. Yes, they sensed they were going to be in danger, but they didn't seem to think that they would be the Number One target."

The Union Pacific depot itself, he said, impressed him and made him feel very small. "I wasn't so much intimidated by the building as fascinated, I suppose. To me it seemed like such a big, stately depot. Afterward, when I got older and went to Omaha and Chicago and saw the train stations there, I knew that the North Platte depot had not been anywhere near the size of those. But when you're young, the first train depot you see makes an impression on you.

"When I would go home at night after helping out at the Canteen, I would remember some of the faces. There were always so many soldiers there, but the faces I would remember were the ones that reminded me of people I knew. I would lie awake."

I watched a few innings of the softball game—the hometown Sensations were having a pretty tough time of it against the girls' squad from St. Louis—and then I said goodbye to Pook German's dad, and to some of the other parents I had met, and headed out to the parking lot.

Near the left-field fence, a man in his seventies—perspiring profusely in the heat of the summer sun that had yet to sink all the way below the horizon of the distant farmland—sold tickets to people who were arriving late for this game, or arriving early for the games that would be played later under the lights. "Have a good time," he said pleasantly to each person who entered the softball complex.

I took a good look at him, and realized who he was:

Lloyd Synovec—the young sailor who had played the piano at the Canteen that day long ago, the seventeen-year-old sailor in the old photo, pounding away at the keys beneath the cartoon of Hitler. At seventy-three he was volunteering out here—spending his time in the sun helping the community by tending the ticket gate.

What was it he had told me about North Platte? That it had never been a big town. But that it seemed to have a good heart.

Eight

I wanted to meet some of the businesspeople—the men and women who made the town run—and I heard about a monthly gathering sponsored by the Chamber of Commerce. Always held on a Friday, after working hours, this get-together was meant to develop and maintain friendships among executives and managers of local companies.

One of these receptions was being held while I was in town, and I stopped by. The first thing that struck me—it made me smile—was the theme of the event. It was a nautical party—a seafaring soiree. Which, in North Platte . . .

Well, there may be nowhere in the United States farther removed—geographically and symbolically—from either ocean than this part of Nebraska. The party itself was being held in the appropriately named Sandhills Convention Center. Yet there they were—waiters and waitresses dressed for sea duty, surfside decor all around the room, fishing nets and boat hulls and island regalia. As long as you didn't look outside the building, you might will yourself into becoming half-seasick.

But as I got to know some of the men and women from the local banks and real estate firms and retail stores, it was not so much the oceans-in-Nebraska aspect I was thinking about—it was the plentitude of the food.

Being offered at serving stations around the room, and by wandering waiters and waitresses, were crab claws, peel-and-eat shrimp, bacon-wrapped scallops, smoked salmon, crab rangoon, chicken empanadas, steamed baby red potatoes with cream cheese and caviar, deviled eggs, make-your-own gyros, Jamaican jerk chicken, fruit tarts. . . .

This was the Chamber of Commerce's way to emphasize the town's self-image of prosperity. But as the food was being served, and lines formed at the multiple bars, I couldn't keep my mind off what I had been learning about the obstacles that had faced the people of North Platte during the days of the Canteen. Because, for all the efforts that were made in the 1940s to feed every soldier and sailor who passed through on the trains, it was done against the backdrop of national rationing. Food was far from unlimited during the war years; groceries, especially certain kinds, were at a premium. Yet the Canteen never missed a day, never missed a train, never allowed a serviceman or servicewoman to go hungry.

At the Chamber of Commerce reception, I watched the guests fill their plates, and I thought about a woman from the Canteen days with whom I had spoken—a woman who explained to me how she and her fellow volunteers at the depot had gotten the job done, all those years ago.

She was Hazel Pierpoint, ninety-one, and she told me that her specialty at the Canteen had been "baking angel food cakes—hundreds and hundreds of them. That's what I did for all those years."

To bake an angel food cake—much less hundreds of them—eggs are required. And although procuring food was not as difficult in west-central Nebraska during the war as it was in some urban areas—much of the food here was produced on the farms—there still was not enough to go around.

Thus, Mrs. Pierpoint told me, she had turned to turkey eggs.

"My cousin over in McCook had a husband who drove

a truck that brought turkey eggs to a hatchery," she said. "When he found out that I was baking angel food cakes for the Canteen, and that I couldn't get enough chicken eggs, he said: 'Why don't you try turkey eggs?'

"Now, I don't know if you're familiar with turkey eggs. They are big. A turkey egg is a like a little kid's football. Like a peewee football. The white is a bit tough. But it will do, for a cake.

"Everyone around this area was a little hard up at the time. Those who had chicken eggs either sold them so they could buy groceries, or traded them for groceries. But with turkey eggs—my cousin's husband gave me the ones from his truck that weren't going to be useful at the hatchery—I could get by.

"It would take me only six or seven of the turkey eggs to make an angel food cake. It would have taken twelve chicken eggs to make the same cake. I was making so many cakes for the boys who came to the Canteen that I always asked my husband to help me beat the eggs. We did it by hand, and all that beating of eggs was wearing him out.

"We didn't have a lot of money—he was a painter at a garage in town—but he wanted to make the egg-beating easier for me, and for him, so he went to Montgomery Ward and got me an electric mixer. I had never had one before.

"The first day I had it in the kitchen and plugged it in, I was beating some eggs with it and the telephone rang. I walked eight feet to pick up the phone, and by the time I turned around the cake batter was all over the kitchen! On the stove, on the table, on the refrigerator . . . I had forgotten to turn the electric mixer off. I didn't curse, but I felt like it."

She said that baking the angel food cakes for the soldiers and sailors had made her feel whole. "I always have baked," she told me. "I learned to bake when I was eight years old. I lost my father in 1919, when I was a young girl. We had a farm in Wellfleet. He had the flu and was not able to recover. My mother took in boarders to support us. The banker, the lumberman, the teachers, some railroad men—I learned how to bake so that we could feed them at our boardinghouse.

"So when the Canteen started, of course I said I would bake those cakes. I fixed as high as eight cakes every week. It didn't cost me much—my turkey eggs were furnished, and so was my sugar. I would use my own cream of tartar, and my own vanilla. I had to make my own frosting—I made it with sugar and milk. We always put 'Happy Birthday' on the cakes, but you couldn't put a name, because you didn't know the names of the boys who would be coming through. Sometimes I put 'Happy Birthday' in red, sometimes I put it in blue—depending on what colors

I had. I used those little vials of coloring, and I mixed them with powdered sugar."

I asked her how she got all those cakes from her house to the Canteen.

"My husband had a car," she said. "We had special boxes to carry the cakes to the Canteen. The ladies at the Canteen would take the boxes out of the backseat of the car, carry them in and lift out the cakes, then bring the boxes back out so I could take them home and make more cakes."

She said she remembered one day especially: "One time I took the cakes down to the depot, and everyone was very busy, and they told my daughter and me just to take the cakes onto the train and see who had a birthday. We were always told that you weren't supposed to go into the part of a train where the boys were sick, and that if you were around sick soldiers, you were supposed to remove your shoes.

"That day, we forgot. We carried our cakes onto a hospital car of the train, and we forgot to take our shoes off. It was just like any other car on a train, except it had beds and equipment to take care of the sick and injured boys. Their legs were propped up, the boys were bandaged up. . . .

"The boys saw us and said, 'Well, you're the first ones to ever come in here.' Which is why we knew we weren't supposed to be there. But they were happy to see us, and they were very happy to be given the cakes."

To this day, she said, she feels proud of what she did during the war. "I could bake a cake, and that's what I could do for the servicemen. They would get letters at the Canteen from soldiers all over the world, thanking everyone for the birthday cakes. I would see letters that said, 'That angel food was delicious.' I knew that was me."

Speaking of thanks, I said, did the ladies at the Canteen thank her every time she dropped those cakes off in the car—did they express their appreciation for what she was doing?

"They thanked me, I thanked them," Mrs. Pierpoint said. "Everybody thanked everybody. We all thanked each other all the time. That's just how it was around there."

She told me that she still baked cakes. Her husband has been dead for thirty years, but at ninety-one she will bake cakes for herself.

"Last week I was going to fix a chocolate one," she said. "I was going to get chocolate and powdered sugar and flour and mix them together. But I was tired. I'm not able to do everything anymore. I was going to make it for me or for my neighbors, but I was just too tired to bake the cake. I will, though.

"Back during the Canteen days, I never got tired. For an angel food cake, you have to take and mix egg whites like you would mix whites for a pie. You have to whip the whites for half an hour. I had a lot of muscle.

"I had a lot of strength, during the war. There was nothing I couldn't do, then."

The willingness of the Canteen volunteers to simply give away food was even more striking in light of the specific restrictions placed on American families by the government during the war years. It wasn't just the self-lessness of the volunteers, although that was impressive enough; it was their selflessness in the face of personal deprivation.

While I was in North Platte I found a ration book that had been issued in 1942. It had been signed for by Irene P. McKain, who was fifty years old at the time and who lived on Rural Route 2. In the book—authorized for Mrs. McKain's use by the U.S. Office of Price Administration—were variously numbered stamps, to be presented to merchants in order for her to be permitted to purchase specific items. The government, on the front of the ration book, warned her of what could happen if she misused it:

"Punishments ranging as high as Ten Years' Imprisonment or $10,000 Fine, or Both, may be imposed under United States Statutes. . . . This book may not be transferred. . . . It may be taken from the holder by the Office of Price Administration."

The instructions were precise:

Each stamp authorizes you to purchase rationed goods in the quantities and times designated. . . . Without the stamps you will be unable to purchase those goods.

Rationing is a vital part of your country's war effort. This book is your Government's guarantee of your fair share of foods made scarce by war, to which the stamps contained herein will be assigned as the need arises.

Any attempt to violate the rules is an effort to deny someone his share and will create hardship and discontent. Such action, like treason, helps the enemy.

Be guided by the rule: "If you don't need it, DON'T BUY IT."

Yet, in the face of this—with food products so scarce— the people of the Canteen kept giving their own food away. From Canteen records I found:

Contributions from the Moorefield group yesterday were 25 birthday cakes, 39½ dozen cup cakes, 149 dozen cookies, 87 fried chickens, 70 dozen eggs, 17½ quarts of salad dressing, 40½ dozen doughnuts, 20 pounds of coffee, 22 quarts of pickles, 22 pounds of butter, 13½ quarts of cream. . . .

Sixteen women of the Paxton community donated 52 dozen Easter eggs, 600 bottles of milk, 2,000 buns,

six hams, 12 sheet cakes . . . one quart of chicken
spread . . . three boxes of apples. . . .

It never stopped—and then, having donated the food
they could have used themselves, they diligently prepared
it so that it would be waiting whenever a train became visible in the distance.

And on those trains?

Riding on those trains were men like U.S. Marine
Sergeant Vincent Anderson, a survivor of the Battle of the
Coral Sea, whom I found living in Palm Desert, California.

"I had never heard of it before we got there," Mr. Anderson, now seventy-nine, told me. "Never heard of North
Platte, never heard of the North Platte Canteen. We were
four cars on a passenger train—four cars full of Marines.
We pulled into North Platte for this very quick stop—and
this lady came up to me and said, 'Is it your birthday?' I
said to her that, no, it was not.

"And she said to me, 'I'm making it your birthday.' And
she handed me this beautiful, home-baked cake.

"I was really . . ." He tried to find the right word.

"I was really *melted*," he told me. "What I saw them
doing at that place melted me. Such kindness. Such kindness."

The first he knew that something good might be waiting for him at the railroad station in Nebraska was during a trip from California for training on the other side of the continent. "There were some MPs traveling with us," he said. "Right before we got to North Platte, they said to us: 'Here's a place you're really going to like.'

"And boy, were they right. All of the guys on our train—they were all so excited. They just thought it was the greatest thing in the world, the way these people treated us. They couldn't believe it. Here were these older women, our mothers' ages—and they made us feel we were heroes. Not all of us were combat veterans—but they made every one of us feel like heroes."

He said that when he came back across the country six weeks later, "I said to myself, 'I'm going to be well prepared and hungry when I get there.' So I didn't eat for twelve hours. I really took advantage of what was waiting for me in North Platte. I was looking forward to it the whole train ride west, and it didn't let me down. I knew it was out there, like a warm spot waiting for me—like a comfort zone."

I asked him what his specific memories of the town were—what made it stay with him all these years later.

"You can tell, when people are being nice to you, if they really mean it," he said. "You can tell it in their eyes—if it's real. If it is, you can see it and you can feel it.

"It was real in that town. And when you talk with servicemen about things you've done, it always comes up. It came up all the time in the South Pacific. 'Were you ever in North Platte, Nebraska?' It was always North Platte—always the one city guys brought up. 'Have you ever been in North Platte?' Because everybody had been there. Everybody had been there on a train.

"I wrote a letter to the town. I think it was after I got back to the base in California, after my second trip through the Canteen. I just wrote it to the postmaster of North Platte. I told him I appreciated how well they had treated us. I didn't hear back, but I don't think I had a permanent address during the war."

He was a platoon sergeant in the Marines; after the war was won, he said, he got a job in claims management for an insurance company. He never went back to North Platte.

"The funny thing is, for all the feelings I have for North Platte, I never saw the town itself. I never got out of the immediate area of the Canteen.

"There may have been a lot of love in the eyes of those ladies who greeted us, back then. But there's a lot of love in *our* eyes, now. Love for that town."

I told him that with all the memories of combat a man like him must have, it was a little surprising that when he remembers the war, those brief minutes at a Nebraska train station would still mean so much.

"I have a lot of combat memories, yes," he said. "But believe me—there's room for North Platte in my memory. There's room for it in all of our memories."

———————

What Mr. Anderson had said about never seeing the town itself—about just seeing the train depot—was something I kept thinking about.

Because if he were to see North Platte today, he might be struck by how far it has moved away from itself, away from the train station. It's not the only American city to experience such a thing—in fact, it's a common tale. But it says quite a bit about what America's cities were then, and what they have become now.

One morning I made a point to start downtown, near the old Canteen, and move gradually outward, toward the interstate. Away from what used to be—away from the place of Mr. Anderson's memory.

Nine

Downtown, had you not looked at a calendar or at the date on the morning *Telegraph,* could have fooled you into thinking you had stepped into 1954.

Around its compact core—Dewey Street, Jeffers Street, Bailey Avenue, Fourth, Fifth and Sixth Streets—it had the hemmed-in, self-contained feel of America's downtowns when those downtowns were the magnets of the cities they centered. Red brick road paving, and solid gray concrete buildings with their dates of construction carved in deep, hollow Roman numerals ("MCMXIV" over the door of the building at 100 East Fifth, the one that contained the city offices of the Metropolitan Life Insurance Company) . . .

One signal that this downtown, like so many of America's downtowns, was perhaps no longer the vibrant heart of the region's commercial life was that the names of the majority of the businesses were purely, almost plaintively, local—you would not recognize them outside of North Platte. In a nation long grown accustomed to chain operations being equated with prosperity and customer familiarity, North Platte's downtown seemed composed mostly of business concerns whose reach did not exceed the next intersection.

Hirschfeld's Fashions for Men, Hoover's Fine Jewelry, Young's Sporting Goods (REMODELING, the hand-lettered sign informed passersby), the Brick Wall Restaurant, the Shepherd's Books and Gifts . . .

It was the downtown of our parents' parents—it was the comforting and unapologetically insular downtown of the soldiers who went off to war, and of the mothers and fathers, sisters and brothers, who waited for them in the years just before television and the interstates changed everything. I walked through this downtown, and I could almost see it as it was in the days when if a person wanted to go shopping, he could find anything in the world he needed, right here.

Brown's Shoe Fit Company, and Sew What Alterations

(the stores with the play-on-words names seemed to have moved in where more traditionally titled merchants had once thrived), Sam Thomas Insurance and North Platte Boot and Shoe Repair and the Dixon Optical Company, with a sign that looked as if Harry and Bess Truman could have been passing through town on the day it went up and strolled right in the door . . .

CLEAN IS KEEN, the legend on a garbage barrel, padlocked to a light pole, announced. One tip-off to how the city, and the nation, had changed was something quite fundamental: the absence of people on the streets. They had moved on—they had taken their business elsewhere.

There was a time, though—the time of the Canteen, which had been located just a block or so from this cozy core—when the people had gravitated to this part of town as if it held the answers to everything.

"I would take my daughter shopping downtown," said Mattie Rumery, ninety-five. "She would always want a certain kind of shoe—a pair of pretty little pink slippers. But they just weren't available, during the war."

Mrs. Rumery first lived in North Platte when her husband was part of the faculty at a University of Nebraska agricultural facility near the town. "It was called the Station, and we women called ourselves the Stationettes.

There were maybe fifteen or seventeen of us. We would all volunteer at the Canteen."

She said that she learned quickly: "When one of those trains came in, the fellows did not walk—they *ran* into the station. The word had gotten around—they knew they would have *good food*. Some country women would bring in their own chickens, but the rest of us had to buy our chickens for frying.

"In the corn season we would always bring roasting ears for the boys to eat. There was no list of what we were supposed to bring—we could bring in everything we liked, if we thought the boys would enjoy it.

"One of the things the boys liked us to get ahold of and make for them was boiled eggs. In the service they were provided with artificial eggs. What we gave them was real eggs. They would crack the shells and gobble them down."

The reward—for her, for the Stationettes, for all of the volunteers she knew—was something deceptively simple: "There was a feeling you were pleasing them," she said. "It wasn't that we felt we were working for them—*they* were working for *us*. We all understood that. We didn't even have to say it."

Which was where her daughter's shoes—or, more correctly, the lack of them—came in.

Mrs. Rumery would take her young daughter to shop

for shoes in downtown North Platte during the Canteen years, and the selection they found was invariably scant. There was a reason for this:

"It was very hard to get shoes," she said. "They had to have good shoes for the boys in the service, so most of the material for shoes went for that. For the boys. You just didn't have the chance to buy fancy shoes—only the basics, and not much of a selection of that. My daughter wanted those pretty slippers so badly, but you didn't have a choice, you just took what the stores had. I had to buy for her a little brown oxford. The people in the store had to pad the shoe all around inside, to fit her little feet—you couldn't always get the right size.

"She would cry sometimes, and I would cry with her, but I would explain: The soldiers had to have the best shoes. That was the way it was, and that was the way it should have been."

The train depot, she said, was the hub of the city back then—the train depot was the point from which everyone entered North Platte, and then walked into downtown to explore the stores, restaurants and hotels.

"The depot had beautiful detail work on the windows, and all around it," she said. "For those of us who were there, it's kind of hard to believe that it's past history. Just that quickly. During my lifetime.

"Sometimes people would get off the train and see our

downtown, and they would decide to settle in. They got jobs here and raised families here and spent their lives here. All because they stepped off the train at the depot, and walked into downtown, and liked what they saw and how it made them feel."

The Mall—that's how it is known, just those two words—must have, in the 1970s, when it was constructed, seemed like the ultimate alternative to the old North Platte downtown.

You can't walk there from downtown—well, you could, but it would take you too long, everyone drives—and when it was built the idea of it, with all the stores under one roof during the brutal Nebraska winters and impossibly humid summers, must have seemed unbeatable. North Platte wasn't alone in having fallen for the promise of the self-contained, single-story mall—during the part of the twentieth century when The Mall was built, almost identical versions of it were going up all across the United States.

What no one seemed to have considered, in those first days of the covered malls, was just how desolate and cold they might feel once they grew older, and the customers took their leave. Today, inside The Mall (with its white-on-brown exterior sign), the first thing I encountered in the wide, unbroken central corridor was an unoccupied kiosk with the printed notification:

SPECIALTY LEASING. YOU COULD BE SELLING YOUR RETAIL PRODUCT OUT OF THIS SPACE TODAY! SPACES AVAILABLE BY MONTH, YEAR, OR FOR THE HOLIDAY SEASON. FOR LEASING INFORMATION, CONTACT THE MAIN OFFICE.

The Mall was closer to the interstate than downtown North Platte was—downtown had been built near the railroad tracks, or maybe it had been the other way around. But whoever had constructed The Mall had done so with the knowledge that the customer base could not be assumed to be near the tracks, but instead near the strip of nonstop highway that reached from one end of Nebraska to the other.

Here were the store names that anyone in any state would recognize: Bath and Body Works, Radio Shack, Dollar General, Foot Locker, Waldenbooks, Payless Shoe Source, General Nutrition Center. Yet, like downtown, there was a permeating silence here. Wherever the people of North Platte were congregating, it wasn't underneath this roof.

At least downtown had those red brick streets; at least downtown felt as if it probably held a place in someone's heart, if you could find the person. The Mall . . .

The stores were open for business in the central part of

the structure, but at the far ends, like at so many of the older malls I had seen across America, there was mostly quiet. Over what was supposed to be the welcoming, wall-less frontage of one store space was a metal gate, pulled down to guard nothing but a vacant interior. NOW LEASING, the sign said. DIAL PROPERTIES, OMAHA, NEBRASKA.

This is what had done away with downtown.

"He's in the hospital, but he wants to talk to you."

The son-in-law of Paul Metro, seventy-eight, of Edison, New Jersey, was letting me know that even though Mr. Metro was facing surgery the next day, he wanted to tell me about his brief time in North Platte. It still meant that much to him.

"I was one of the men those people went out of their way to be nice to. . . ." Mr. Metro began when I reached him in his hospital room.

He started to cry softly in mid-sentence. I would have attributed it to the stress of being about to go into surgery, except that it was happening regularly when I spoke with the men who had come through North Platte on the trains. The volunteers from the Canteen, while emotional, usually remained composed. But the soldiers they had welcomed . . . as often as not, they would weep at some point during our conversations as they recalled the experience.

"It was a furlough from Wendover, over Christmas," he

said. He had been assigned to the unit that prepared the B-29s for the eventual flights that would end World War II; he had been a radar mechanic based in Wendover, Utah, under the command of Colonel Paul W. Tibbets, who would fly the atomic bomb to Hiroshima.

"I was heading home to New Jersey for Christmas," he said. "The conductor told us that we would be stopping in a town where there was a free Canteen. I had not seen much of the world before going into the service. I went to high school in Linden, New Jersey, and the farthest west I had ever been before the service was when my father took us on a vacation to Pennsylvania once. On old U.S. 22—a two-lane highway through little towns. That was my idea of travel."

Training in the remote salt flats of Utah for the atomic bomb mission, he and his unit operated under almost total secrecy—even their families were to be told nothing about what they were doing. So as he progressed toward his mother and father's house in New Jersey for Christmas . . .

"We stopped at the station in North Platte, and a mob of guys were running toward the depot," he said. "Everybody sprinted, like the place was going to run out of food. But there was plenty for everybody. I had a sandwich and some milk, and I spoke to some of the people who greeted us, just thanking them for doing it.

"Those people . . ." He began to cry again.

"Those people were working so hard to show their

concern and regard for all the servicemen. I think if more of us had realized just what an effort those people were putting out for us, I think maybe we would have sent more thank-you letters to them, so that they would have known how much we appreciated it. I worry sometimes that they didn't know—we were there and gone so fast."

He talked about shoes—I hadn't said a word to him about what Mattie Rumery had told me about her daughter and the shoe stores with the limited selections, but he brought it up himself.

"Do you know what all the civilians did during the war?" Mr. Metro said. "We in the armed forces get all the credit for fighting the war—but the way they rationed things, so that we could have what we needed . . . We always had good pairs of shoes to wear all the time, without ever having to worry about it—we had good shoes because the civilians were doing without. We got the best."

He said that of the twenty-seven men in his radar unit, only five were still alive. He was speaking to me from the Robert Wood Johnson Hospital in New Brunswick, New Jersey, he said; he told me he was divorced and lived by himself. He was scheduled to have his gallbladder removed in the morning. He said he had spent his time after the service as an insurance field man; he had already undergone heart bypass surgery, and had been diagnosed with prostate cancer. When he had begun to feel the pains that would necessitate this operation, he said, there was no one in his house for him to talk about it with; he had called 911, and here he was.

He said he was feeling all right, although a little scared. "They gave me a shot of that drug they used to have in the Army—what is it? Morphine."

I asked him why he had sent word that even on a night like this, he was willing to talk with me.

"I know that many of those people in North Platte are older now, like me—or they're gone," he said. "When I think of them, I think of young girls handing out magazines and fruit at the windows of the trains. But of course they're not young girls anymore.

"They'll all be gone soon. All of us will be gone. And I think America should remember those people. Right in the middle of the country, with all those trains going east and west—railroad transportation was really the only thing at the time. Those people in that town—they helped us. They made us feel that someone appreciated us."

He broke down again.

"I'm sorry," he said. "It's a lonely day here."

On the farthest outskirts of town, I found it.

The place—and the people.

Here was what downtown had evolved into. Here was

the focal point of the city's commercial life—here was its heartbeat.

The Wal-Mart SuperCenter. Open twenty-four hours. Two minutes by car from the interstate exit.

The parking lot was so full when I arrived that it was a challenge finding a space. The moment I walked in I found everything that downtown wasn't, everything The Mall must once have aspired to be.

The store was packed with people—you couldn't see from one end of the big building to the other, of course, but the brightness of the lighting, the excited murmur of the shoppers, the country music playing from ceiling speakers (Wal-Mart knows the tastes of its local communities) . . .

This was it—this was where the town had moved. Never mind that there was a Wal-Mart like this seemingly in every community all across the country; never mind that, unlike the old downtowns, the buying decisions about the products that would stock the shelves were made not at the local stores, but at a central headquarters. The fact was, those shelves were overflowing. Employees were everywhere, keeping things in order, offering to help; the Wal-Mart greeters were at the front doors, telling the nonstop streams of arriving customers that their presence was appreciated.

A massive grocery with fresh fruit and vegetables, the

aroma wafting through the aisles . . . an eyeglasses department with an on-site optometrist . . . a barbershop . . . a
school-supply department . . . a full-service pharmacy . . .
luggage, and carpeting, and television sets, and computers,
and a bank, and kitchenware, and a portrait studio . . .

It wasn't that the Wal-Mart felt distant and national
and corporate-controlled; that was the usual criticism
about Wal-Marts, but if the SuperCenter had felt like that,
the people of North Platte would not have been jamming
the store. No, what the Wal-Mart near the North Platte
exit of I-80 felt like was what the old downtown must
once have been.

It felt like . . .

What an odd and paradoxical word to use, for a place
like this . . .

It felt like home.

Wherever a person was from, this felt like home, or
what the American idea of home has evolved into. This felt
like hometown shopping. The real place—downtown
North Platte—didn't feel like the real place. This did. All
these different stores, inside the Wal-Mart—except they
weren't different stores, they were the same store. And if
Wal-Mart had stores like this all across the United
States . . .

Well, that only helped. Because to the people who
were shopping, no matter where they were from, the store
felt like their town. They'd shopped here before. Whatever

brought them here today, they were somewhere they'd been. It seemed familiar—much more so than those silent brick streets near the tracks downtown, much more so than that chilly and barren mall. This was where they could count on feeling at ease. They'd been here all their lives.

The city had moved away from the city. In the blocks around the Wal-Mart were the other places that made people sense they were at home—the Hampton Inn, Applebee's, the Quality Inn where I had my room. Out here, North Platte had moved away from North Platte, and had been pulled, as if against its will, toward America.

And if somehow the soldiers of World War II were to come through again—if somehow the troop trains, on their ten-minute stops, were to steam in here—would a town do now what this town did for those six million soldiers sixty years before?

Inside the Wal-Mart, black-colored half globes mounted at measured intervals all along the ceiling—black-colored half globes of the kind you see on casino ceilings, the globes covering surveillance cameras—stared tirelessly down upon the throngs of shoppers. The railroad trains of the Canteen days, and the trust of those Canteen days, were far away.

Would a town today do what this town did then?

The question, today, is more elemental than that.

The question, out here by the exit ramp, is:

What's a town?

Ten

Having immersed myself in the more wholesome facets of North Platte, I decided, in the interest of balanced reporting, to seek some vice. It was not going to be easy.

I knew there were no casinos in town—I had asked, and had been told there were none anywhere in the area. No strip clubs, either, or lap-dance emporiums—not a one. Nothing in the Yellow Pages to indicate the town even had a seamy side.

Which is why—in the company of two nurses from the local hospital, who had volunteered to be my chaperones—I found myself on my way to the weekly bikini contest held at the Touchdown Club. This was likely to be as close as I would get to the racier precincts of the town. At least as the town was now.

Such precincts were not always so hard to find—which is why I was looking. In the years before the Canteen, I had been told, Front Street near the railroad tracks had been lined with brothels. Apparently the town had quite a past. As railroad historian George H. Douglas has written:

> When the winter season came on in 1866, the Union Pacific had already passed the long-awaited 100th meridian—247 miles west of Omaha—and gotten to a spot 43 miles farther west. They called this place North Platte, and decided to settle down for the most savage part of winter.
>
> Tents and shacks were put up, and because North Platte was the temporary end of the rail line it became a brisk and often violent starting off point for overland traffic to the West. . . . Gathering about were every manner of Indian fighter, gold seeker, homesteader, and roustabout. . . .
>
> In a matter of only a few weeks a hundred jerry-built structures had blighted the area—saloons, bordellos, hotels, warehouses, stables. The huge army of boisterous workers was inevitably followed by the usual assortment of gamblers, shoot-em-up boys, whiskey

salesmen, and miscellaneous troublemakers. A correspondent for the *New York Times* deftly characterized the atmosphere:

"The largest part of the floating population is made up of desperados who spend their time in gambling of all kinds, from cards to keno to faro. Day and night the saloons are in full blast, and sums of money varying from five dollars to fifty and even one hundred change hands with a rapidity astonishing to one who is not accustomed to the recklessness which their wild frontier life invariably begets."

A reporter for the *Missouri Democrat,* Henry Morton Stanley, dropped in on the scene in early Spring when construction was about to begin again. "Every gambler in the Union," he wrote, "seems to have steered his course for North Platte. . . . Every house is a saloon, and every saloon is a gambling den. Revolvers are in great requisition. Beardless youths imitate to the life the peculiar swagger of the devil-may-care bullwacker and blackleg."

But then came the war, and the Canteen.

"I wasn't old enough to work in the Canteen, even as a platform girl," said Doris Dotson, seventy-one, who still lives in North Platte. "I was only twelve or thirteen during the first years of the war.

"They had a jukebox in the Canteen as well as the piano. My cousins had taught me how to jitterbug. I would go down there to dance. So many of the soldiers on the trains were boys not that much older than I was, really. You know that expression, 'I'd rather dance than eat'? The boys would pass up the food tables to come down and dance with my friends and me."

She said that the ten-minute stops the trains made at the depot sometimes stretched to twenty minutes, but not much more than that: "As long as it took to put water in the steam engines." The atmosphere, she said, was "wall-to-wall—like a big cocktail party—shoulder-to-shoulder."

As soon as she and her friends heard music beginning to play, "Us girls started acting silly. We started dancing with each other. The boys would come over and we'd put our hands out, like we wanted to dance. No slow songs— we didn't want to make the boys homesick for their girls. We didn't want to make them sad. 'In the Mood,' 'Tuxedo Junction' . . . anything that was fast, we'd dance to."

Although she was spending so much time dancing with the soldiers and sailors at the depot, "My mother never worried about me. As far as I was concerned, I was being very patriotic.

"I had a jacket that I called my insignia jacket. I just started collecting patches from the units of all the soldiers I danced with, and I sewed them onto my jacket. If they didn't have the patches with them, I carried little slips of

paper with my address on them. It said 'Doris Hedrick'—
that was my name then—'402 South Chestnut, North
Platte, Nebraska.'

"The soldiers who would dance with me would always
look at my jacket and say, 'You don't have the best patch
on there.' A few days later, I'd get a letter with their division's or platoon's patch in it. My dad finally painted our
mailbox red. He said we had the hottest mailbox in
town."

Her family lived about ten blocks from the station, she
said: "You could walk down there after school every day,
around four o'clock, and you'd know that you would catch
a train every hour. A lot of times there were at least two
trains an hour."

As much fun as she was having at the Canteen, she said,
"That was also kind of a sad period in my life. Not just
mine, but everyone else's. I got a job working as a carhop
at a drive-in restaurant—the Shady Inn, at A and Jeffers.
Hamburgers, Cokes, malts. Beer, too—you could take beer
out to the cars. You had to be twenty-one to drink it, but
if you were a carhop, you could carry it to the cars no matter how old you were.

"I was still going down to the Canteen to dance. By
the time I was sixteen, I had graduated to being a soda jerk
at the O'Connor Drugstore at Fourth and Dewey. And
during all those years, all those troop trains kept coming
through, day and night. . . .

"What I meant about being sad is this. I'd dance with some soldiers, and they'd send me a patch. I'd write them, and they'd write back, and I'd keep writing to them as long as they wrote back. But then, when their letters stopped . . .

"Did they just get tired of writing? Or were they killed in action? I never had any way of knowing. I didn't dwell. I hoped that they had just gotten tired of writing to me, instead of the other."

She doesn't dance anymore.

"I had a stroke," she said. "I'm down to my cane and my walker. I don't think I could jitterbug.

"But I still have my insignia jacket. All those patches. All those boys."

From the street one afternoon—South Jeffers Street—I had seen the sign advertising the bikini contest. The sign said there was a $1,500 first prize, and that the contest would take place on a Friday night inside the establishment where the sign was posted: the Touchdown Club.

There seemed to be no hint in town of the pre-Canteen North Platte—none of the seedy enterprises that were alleged to have once made the city tawdry. So I concluded that this would have to do. I had no idea what the bikini contest entailed, but I knew, as a visitor in a small town, I probably shouldn't be seen skulking around the place alone. I mentioned my misgivings to two nurses who worked at the hospital that was along the route of my daily morning walk—the Great Plains Regional Medical Center—and they said that if it would make me feel more at ease, they would accompany me.

They asked their husbands to watch their children, and just as full darkness was falling, the three of us walked into the site of the bikini contest. It was . . .

Well, it was a sports bar. A sports bar with a University of Nebraska theme. Photos of football players all around, Husker memorabilia . . . there seemed to be a stage of sorts in the front of the main room, but there was no one standing on it.

We took a table, and ordered some beers. The place was mostly empty, except for a table in the restaurant section, at which was seated a girls' softball team, with parents in charge. We talked, and waited, and talked, and waited, and talked . . . and nothing happened.

One of the nurses told me that she recognized a female employee of the Touchdown Club—she had helped deliver the woman's baby at the hospital. She asked the woman what was going on.

The woman said they didn't want to start the bikini contest until the softball girls had finished eating, and were off the premises so they didn't have to witness it. It wouldn't be right.

Very thoughtful. The softball girls took their time, and when they finally departed, bats and gloves in hand, the lights went down and big-beat music started to course through the room.

High-beam colored lights swept the stage. It was time.

And no one came forward.

There were no entrants.

I looked around. Customers sat at two or three of the other tables in the bar, but that was it. And the bikini-wearers? The women who would strut about in pursuit of the $1,500?

There were none.

This was vice in North Platte.

The nurses and I looked at one another. We looked at the empty stage.

"This may be a long wait," one of them said.

Larry McWilliams was a young boy in North Platte before the Canteen turned Front Street into a chaste, patriotic oasis. He's seventy-one now, living in Golden, Colorado, and he told me that he smiles when he thinks of the contrast between the North Platte of the Canteen days, and the North Platte of the years just prior to that.

"There were lots of bodies to be found in the cornfields," he said. "North Platte was known as Little Chicago, back then. Very rough and wide-open. One of the brothels was known as the Como Rooms. The most famous madam in town was Dotty LaRue.

"Patriotic? I think North Platte considered itself to be patriotic even before the Canteen. The people who ran the town appreciated their freedom—the freedom to be corrupt."

Nevertheless, he said, it was a safe town for a boy to grow up in—and when the Canteen opened, the entire atmosphere around the tracks changed. "I'd usually go down to the Canteen with my buddies at least once a week. I was eleven years old. My parents would never worry about me wandering around the tracks—anyone let their kids do that when the Canteen was up.

"The place was absolutely humming, all the time. The girls from North Platte High School were always there. It gave them the opportunity to flirt on a short-term basis. To be noncommittal. After all, the train was going to be leaving in twenty minutes. Sometimes there would be a little smooching going on. To raise the morale of the soldiers they'd just met."

And it wasn't just the young girls who kept their eyes open when the trains came in, he said. "I heard my Aunt Regina Rector telling some women once that Fred Astaire was at the Canteen, coming through on a special services train that carried movie stars. She said that Fred Astaire started dancing with her on the platform."

The arrivals of the troop trains were so constant, Mr. McWilliams said, that in the wintertime the snow around the depot turned black: "There were so many trains, with so much soot from the engines, that the soot just became a part of the snow on Front Street. That's one of the memories of the place that stays with me: black snow."

He said he and his friends loved the Canteen, but they hadn't minded the North Platte that had preceded it, either.

"We were so young then, we weren't really aware of what was going on in the town. We would be at the soda fountain at the corner drugstore, and the prostitutes would come in for a cherry Coke, and we knew something was up. We just weren't quite sure what it was."

At the Touchdown Club, forty-five more minutes had passed, and still there were no women in bikinis on the stage. Management appeared to be getting nervous.

A burly fellow who worked for the place began to circulate through the crowd, what there was of it. He held a yellow legal pad in his hand, and was approaching tables. I saw that he was inviting women customers to enter the contest. He was almost pleading. He was getting no takers.

"Fifteen hundred dollars," I heard him beseech one young woman who was having a burger and a drink with her date. She shook her head no.

This was better than anything I could have imagined. The one place in town openly devoted to something a little off-color . . . and the stage sat bare.

"We're really sorry about this," one of the nurses said to me, as if they were letting me down.

"Sorry?" I said. "This is great."

The fellow who worked for the bar came up to our table with his legal pad. He said to the two women: "Can I interest you in entering our bikini contest? We still have spots available."

One of the nurses, who was forty-one years old, laughed out loud.

"I'd love to," she said, "but I have to get home to my family."

He went on his way. The music boomed. The lights crisscrossed the empty stage.

What was that line I had read, in the old newspaper story from the days of the rough-and-tumble railroad gangs? "The peculiar swagger of the devil-may-care bullwacker and blackleg"? What was it that Larry McWilliams had told me about his North Platte boyhood? About the prostitutes coming into the drugstore soda fountain for their soft drinks?

This was going to have to do. There was a bowling alley right next door—actually, attached to the Touch-down Club. This was going to have to suffice for my big night out.

There are nights out . . . and then there are nights out.

"It had to be in 1944," said William F. Kertz, seventy-seven, of Elmwood Park, Illinois. "I was on my way to the West Coast on a troop train. In the Army during the war I was in England, and France, and Belgium—but I know that the train ride was in 1944.

"I didn't get to see the town of North Platte—just the station. I got off the train, and they were very generous to us. It made me know what Nebraska was like."

There are nights out. . . .

"At a later date," he said, "I was stationed with the Army outside Omaha. I went to a dance at the Fontenelle Hotel. I met a girl. Such a pretty girl. Her name was Shirley."

Shirley Gilstad, to be exact. "I was working for Metropolitan Utilities," Shirley Gilstad—now Shirley Kertz—told me. "My girlfriends and I went to a dance one night. It was kind of a shock to have one of the girls say to me, 'There's a soldier who would like to meet you.'

"He said to me, 'I'm Bill Kertz from Chicago, and I'd like to dance with you.' He was very friendly, and very nice looking. And he loved to dance. We danced that whole night."

They have been married for fifty-five years. "He has often said what a nice thing it was for those women to be out at the train station in North Platte when the troop trains went through," Mrs. Kertz said. "But I tell him, he should have figured it out at the dance at the Fontenelle Hotel:

"Nebraska people are like that."

One of the nurses dropped me at the front door of my hotel.

"We could have waited around the bar to see if anyone entered the contest," she said. "But I really have to get home before my kids go to bed."

"Thanks anyway," I said. "But no one was going to enter."

"I know," she said.

"I appreciate the ride," I said.

"Do you think anyone ever enters the bikini contest?" she asked me.

"I wouldn't have any idea," I said. "You live here."

"I know," she said. "But it's the first one I've been to."

"Fifteen hundred dollars," I said. "I don't see how they can afford that, with so few people in the place."

"Maybe that's how they can afford it," she said. "They know that no one's going to enter, so they won't have to pay it out."

She drove off, across the bridge, over the river. I thought of Doris Dotson doing the jitterbug to "Tuxedo Junction," and soldiers being given a smooch to keep their morale up, and black snow by the railroad tracks all winter long.

Eleven

There were islands in the South Platte River—little half-protruding grass islands that dotted the water.

The river itself didn't look all that inviting. Some days I would walk across the bridge that spanned it, other days I would travel a narrow dirt path down to its banks so I could walk at river level. Things were more than moderately grungy down there; oversized and overly used cardboard boxes, and the remnants of campfires, made it apparent that some people were living along the river, making do. It was not a place you would recommend anyone visit after dark.

But during the daytime the grass islands provided a site for children and their parents to play and relax—the hotter the Nebraska days became, the more families I saw sitting on the islands. From down there in the river, the perspective on the world at large was kind of Tom Sawyerish—you had to be up at bridge level, on Jeffers Street, to realize you were really just a few steps away from the Wal-Mart.

I would see the river at the beginning of my walk, and then again at the end, as I returned to the Quality Inn. I was beginning to look forward to all of this, to count on that daily walk through North Platte. It was starting to feel like a place I knew.

One person I kept thinking about was Ethel Butolph's younger sister. Mrs. Butolph, who had met her husband because of the popcorn ball at the Canteen, was a woman replete with the aura of romance—when I had spoken with her, I almost could close my eyes and see her courtship with the soldier who fell in love with her.

But, of course, Mrs. Butolph's name had not even been in a popcorn ball—the whole point of her story was that it was her sister Vera whose name had been put into the ball, and because of this, and where it had led, Ethel met her husband-to-be. Sort of a three-cushion shot into the side pocket.

Still, it had worked. I wanted to speak with Vera to hear the tale from her side.

I heard it, all right.

"I was just going to high school," Vera Butrick, now seventy-seven, told me. "The club of ladies from Tryon went down to the Canteen in North Platte when they were scheduled for their duty there, and they were the ones who put the names in the popcorn balls. We didn't even know they were doing it.

"So one day I received a letter from some guy— William Woodrow Butrick. The letter was delivered to Tryon High School, not to our house. He said in the letter that he was from White Cloud, Kansas.

"He just introduced himself in the letter. He said he was a farm boy, and that the ladies must have put my name in the ball. So I wrote him back. A little cheering-up letter. I was a farm girl, too."

She told me she was one of several girls in her high school class who received letters from soldiers. "Some of the girls answered, some didn't," she said. "I did. I wrote to Woody, and that's how it happened that he said he had a friend who needed a pen pal, and my sister got involved with the fellow she would end up marrying."

Vera and Woodrow wrote back and forth for quite a while, she said. She had gotten a job in North Platte: "I was working at a laundry, Gambs Laundry, folding sheets. It was boring, but you had to do something." That was when she first met Woodrow Butrick in person.

"He had a furlough and he came to town," she said. "I was living with two of my sisters in an apartment. There was a knock on the door. It was a soldier. It was about eight o'clock in the morning. I was already up, and so was my sister Opal.

"Oh, gosh. I wasn't much impressed. I didn't think that much of him, standing at the door. We went to breakfast at a little restaurant a half a block away. I was so darn bashful it was pitiful.

"He was a big eater. He ate, and we sat there and talked, and then we walked in the park. We caught a ride out to my folks' farm, northeast of Tryon.

"Then he had to go home to Kansas, and after that they shipped him to Germany and France. We kept writing, and the letters became very friendly. Maybe a little mushy, but not much mushy.

"I wasn't really falling in love. He was just nice and everything. I think he got serious first."

When the war ended, she said, she was teaching at a small grade school in Tryon. I told her that this sounded like a pleasant way for a young woman to make a living.

"I thought it was an awful job," she said. "There were only three kids in the whole school. It was just a little country grade school. You could teach in it if you had your high school diploma. Or even if you got a teacher's certificate by taking a test in high school.

"Three kids—that's it, that was the school. A kindergartener, a first-grader and a third-grader. One of the three was my nephew, and the other two were neighbor kids.

When you were the teacher at that school, you did everything. Keep the fire going in the wintertime—whatever needed getting done, it was up to you to do it.

"I did that for two years. I would take marriage any day over that. After the war, Woody came back. We got married, and were married for forty-six years."

They lived on a farm in Robinson, Kansas, she said, until he decided to get a job as a freight handler in North Platte. "We were kind of hard up in Kansas," she said. The job he took in North Platte enabled him to support her, and for them to raise two daughters. She was always grateful, she said, that she got out of that little schoolhouse in Tryon.

"Woody and my daughters were the high point of my life," she said. "I'm not the bravest person in the world. I don't know who I would have met. I'm thankful that someone put my name in the popcorn ball at the Canteen, because that's what did it. For me, and for my sister Ethel, too. Neither of us would have met our husbands otherwise, and our children never would have been born.

"Woody died in 1992. He was a very good man."

Something I kept going back to during my time in town was the guest book—actually, a collection of them.

They were not elaborate. They were stiff-backed bound volumes, in which the soldiers who came through on the trains evidently had been invited to sign their names and list their hometowns. These were the kind of books you sometimes encounter at weddings, set out for the guests to inscribe so the bride and groom and their parents can in future years look back on the ceremony and have a record of who was in attendance.

You also come across these books at funeral homes, so that the family will know who was present to pay last respects. It was impossible for me to keep that connection out of my mind as I leafed through the pages and read the names. I knew that some of these men undoubtedly never returned from the war.

Their penmanship was careful. The soldiers had arrived for their brief time in North Platte during the era of fountain pens, and the signatures had that distinctive look. One name on top of another, for long page after long page, book after book. Those ten minutes in the train station must have speeded by in a blur—it was surprising to me that with all of the food and beverages, all the singing at the piano, all the welcoming greetings from the Canteen volunteers, the men would take the time to stop and sign the guest books.

But they did. Private Harry E. Benson, Hartford, Connecticut; Private Arthur G. Hannel Jr., Buffalo, New York; Private Bill Regan, Philadelphia, Pennsylvania; Private Paul W. Font, Rome, New York; Private Bob Quick, Cleveland, Ohio; Private First Class Joseph F. Snodgrass,

Newtown, Indiana; Sergeant D. G. Christensen, Los Angeles, California; Private Bert T. Hames, Akron, Ohio . . .

They must have been in such a rush. Yet they signed in. It was almost as if they wanted someone someday to know they were once here, if only for those few minutes.

And it had worked. That's what I kept thinking as I looked through the books and moved my eyes closer to all those signatures. All these years later, someone knew.

What I could not know, of course, was what happened to all of those men who had signed in during the war. But once in a while, I was able to find one of them—the bearer of a name in the guest book. Kelly Pagano, for one.

"I was drafted," said Mr. Pagano, eighty-nine. He lives in Waukesha, Wisconsin; he grew up in the Wisconsin town of Marblehead, "a quarry town—they broke stone. My dad worked there in the kilns, for the Western Lime and Stone Company."

In May of 1943, he said, "I was going on thirty-one. I was working in a defense plant, welding oxygen tanks for bombers. The older guys were getting drafted at that point. I was married. I thought it was about time I went in."

He served with the Navy in the South Pacific. On one trip across the United States, the troop train stopped in Nebraska.

"I didn't know what it was," he said. "I asked some of

the other guys. They said, 'This is *North Platte.* There are people in the depot with food baskets for you.' I said to myself, 'Oh, my, this is really something.'"

While he was in the Canteen, he said, "I signed a register. Just to say I was there."

He got back on the train, he continued on toward his Navy service, and when the war ended he moved back to Wisconsin and resumed his work as a welder and raised a family. In the 1960s, Mr. Pagano, his wife and his son and daughter planned a summer vacation in the western United States.

"We were going to Colorado Springs," he said. "It was a driving trip. I had a new Oldsmobile. I had told my son Dick about the North Platte Canteen, and about how I had signed a register book there, back in the forties, during the war.

"On the way back from Colorado, I decided to stop in North Platte. I had not been there since that day. Dick was fourteen, my daughter was seventeen. I said to them, 'My name is registered in that town.' I wanted to show my children how I was treated.

"We found it—after all those years, we found North Platte. We had a map in the car, and Dick read the map, telling me what road to get on. We found the building where they had the old registers.

"And the kids found my name in there! My name, and the date I had been in the Canteen. We had to look for a while, in all the books, and Dick helped me—he was pretty proud that I had been a serviceman.

"After so many years . . . to see my name, the way I had signed it . . . it was quite a moment for the Pagano family. We found a motel and stayed overnight, and we went out to a restaurant for dinner. I kept explaining to my family: You got treated so good by the people in this town. You're on the train, you're leaving for the service, you don't know where you're going. And you got treated so good."

I found Mr. Pagano's son, who is now fifty-three years old and a corporate attorney. "I remember my father being so pleased in North Platte, because seeing his name in that guest register brought back so many memories for him," Dick Pagano told me. "I was fourteen—when you're fourteen, you don't really know what things mean. But I could tell that my father was moved.

"There was a swimming pool at the motel—that's what I remember. That's what's important to you when you're that age. But we had driven all that way, and my father was happy."

Kelly Pagano said he has been married for sixty-one years, and that he finds himself thinking more and more about the way the world was when he set off from Wisconsin to fight the war. "We used to keep the key to our house in the mailbox," he said. "You could sleep in your house with the doors open. Now there are bars on the windows."

But one thing that is unchanged are his thoughts about the Canteen—not just his thoughts of the day when as a sailor he first visited it, but also the day he and his family drove into North Platte in search of something.

"The kids wanted to find my name in that book, and they found it," he said. "There was their dad's name, right where I said it would be."

He asked me to do something for him:

"You tell those people who live there that there are a lot of guys like me who don't know a whole lot about their town, but we know the most important thing:

"We know that it was a place where when we needed it the most, we were treated wonderful."

I would wake up early and watch "The Star-Spangled Banner."

The local television station played it. North Platte is one of the smallest towns in the United States to have its own television station—an NBC affiliate—and it begins each broadcast day with a tape of a choir singing the national anthem.

This used to be somewhat common for television stations. But in many cities, especially the more populous ones, the every-morning playing of "The Star-Spangled Banner" faded away at approximately same rate as the concept of "beginning the broadcast day." The broadcast day, in most places, never stops, so it never starts—it is perpetual.

In North Platte, it stops, sometime between midnight and dawn. And when it starts again . . .

There is that choir. And just in case some other station somewhere wants to boast that it, too, plays "The Star-Spangled Banner" each sunrise. . . .

First verse *and* second?

You get both in North Platte.

Just the sound to put a visitor in the proper mood as he heads out the door for his daily rendezvous with the river and its islands of grass.

Twelve

NO HORSES ALLOWED.

Every time I would see that sign—it was on the edge of Centennial Park—I would start to smile, as if the sign were some quaint gimmick. As if it were a version of those 1950s drive-in restaurant knockoffs you find in a lot of large cities—the newly opened restaurants that strive to remind customers of a previous era, with waiters and waitresses carefully trained to snap their chewing gum and call the patrons "Sweetheart." It's all done with a wink—everyone is pretending. It's theater, stagecraft. Everyone involved—waiters, waitresses, diners—might as well be acting in a play. What is being sold is not hamburgers—

what is being sold is an experience, or the memory of one.

That's what I first thought the "No Horses Allowed" sign was—a determinedly cute reference to times past, a reminder of what the town used to be. A gentle little joke.

But it wasn't—what it meant was what it said: that you weren't allowed to ride or lead your horse into the park. With the interstate highway and the cable-television hookups and the brand-name fast-food restaurants, there were moments when it was easy to half-believe that North Platte was just another interchangeable part of a bland and homogenized America in which Connecticut is no different from Texas, which is no different from Oregon, which is no different from Georgia. All the same, in the ways that matter.

That is the notion we have come to accept about the United States. But west-central Nebraska is still, at its kernel, what it was; west-central Nebraska is spiritually often just a blink away from the place the pioneers first crossed so long ago. NO HORSES ALLOWED at Centennial Park meant no horses allowed—sort of like check your guns at the door of the saloon. In fact, I was seeing a horse every day, right in a city neighborhood. I didn't know what it was doing there, near the intersection of Cornhusker Circle and McDonald Road. It roamed inside a low fence, and some mornings I would see a young woman pull up in a Honda Civic, get out and give the horse some fresh water from a garden hose, make certain the horse had enough to eat, and then be on her way, apparently to work.

That was the only horse I saw in this part of town—maybe the sign in the public park applied to this horse specifically—but the horse, and the admonition on the sign, were a reminder that the people who lived here in the Canteen days must not have had much trouble understanding that they weren't in Indiana or New Jersey. No widespread homogenization of the nation, not back then—this was a place that back then, I sensed, felt like itself, not like everywhere else.

"My dad was our alarm clock," said Waneita Schomer, seventy-nine. "We didn't need any other alarm clock. He never slept a minute past the time he meant to wake up."

Mrs. Schomer might be just about the model witness from an era in which this part of Nebraska could never have been mistaken for anywhere else. The specificity of her memories, the exactitude of her accounts of what the Canteen days were like for her family . . .

"I was born in a little town thirteen miles east of North Platte," she said. "Maxwell. It was in an area right between the North Platte River and the South Platte River. My mother's group during the war was the Valley Extension Club, and they would be assigned to the Canteen at least once a month, sometimes more.

"Everything we cooked or ate on our farm was sweetened with honey. We raised the bees, for our own use. We used the honey from the combs as our sweetener for the apples we canned, for our coffee, even for our cereal. If we had oatmeal, we put honey in it. The only thing I can't stand to eat today is a honey cookie. I just had too much honey when I was young.

"But for the things my mother baked for the Canteen, we bought sugar. My mother and father said the servicemen should come first, so we used very little sugar in our home—we used the honey from our bees—and we purchased the sugar to use for the servicemen."

She remembers in scrupulous detail what it was like on the farm on the nights before a trip to the Canteen. "Late in the day, we would dress chickens. We had no refrigerator, so we had to get the chickens ready for the Canteen just the afternoon before. My dad was really handy with an ax.

"Mother had a great big iron kettle on a tripod out in the yard. We would get it boiling hot. There was no electricity or gas on the farm. We used lots and lots of wood for the fire—and corncobs to start it. We would pluck the chickens. You just dipped them in the water—you didn't hold them in there long. Then we would wash them real, real good. We'd put some baking soda in some water and clean them. Then we would soak them in salt water until two-thirty in the morning.

"We would get three or four hours' sleep. My dad

always knew when to get up—he never slept through. Even if he had to meet someone on a train, he always knew the time in his head.

"We would get up and take the chickens to the Canteen. People are so fussy about refrigeration today. We would just get a bushel basket and line it with oilcloth. Then we would put the chickens in there, and cover them with another piece of oilcloth. We would leave the farm about five-thirty A.M.

"In the car we would have the chickens, and three or four dozen hard-boiled eggs. My dad had an old Model A—a 1928 Ford. It was a big deal for us to drive somewhere—we walked even to church, which was three and a half miles. It wasn't so much to save gas—people think that the gas rationing was because there was a gas shortage, but it was really to conserve the tires. America got its rubber from Japan, and the government didn't want us to burn up the tires. That's why people weren't supposed to just drive around.

"But we would drive to North Platte, to the Canteen. And when the trains rolled in . . . man, the sight of all those guys jumping off . . .

"None of us girls were allowed on the trains. Those guys had been away from women for a long time. We knew we should stay away from the train cars."

I told her that I had heard the soldiers were always perfect gentlemen to the Canteen women.

"Like I say, they had been away from women for a while," Mrs. Schomer said. "Let me put it this way: My mom and dad would have killed me if I had put my name in a popcorn ball."

Her life, she said, had not turned out exactly the way she might have dreamed. "First I worked in a dime store," she said. "F. W. Woolworth. I was getting ready to go to the teachers' college in Kearney, but I had major female surgery two weeks before. And then that was the year the crops failed."

She got married when she was twenty-one, she said, "but twelve and a half years later my husband died. I was a widow at thirty-three." To support herself she went to work for the Union Pacific Railroad, as a clerk, and stayed with the company for thirty-three years.

"I feel very sad that the passenger trains don't come through anymore," she said. "It used to be that if you wanted to go to a big town like Denver or Omaha, you could just jump on the train and go. I could get a train to Denver at five-thirty or six in the morning, shop for three or four hours, and be home that night.

"It was fascinating just to have the trains coming through all the time. If you didn't have a lot of money to spend, you could go down by the tracks and watch the trains, and play pump-pump-pullaway with your friends—where you pull each other apart. Or you could play andy-andy-over, where you would throw a ball over a building and have your friend try to catch it on the other side."

She can still see the old depot, she said, even though it is gone. "Do you know, I can see the soldiers running to make the trains," she said. "In my mind, I can see all of it. I visualize that, more than I visualize the soldiers running *off* the train and *into* the depot. Because there was something about how much they liked being here—they *knew* they had to get to that train or it would leave them, but they always stayed to the very last minute.

"That's what I see, when I close my eyes—those soldiers on Front Street, *running*."

It wasn't that the East Coast, and what it represented, was foreign or exotic to west-central Nebraska during the Canteen days. Then, as now, national companies did everything they could to drum up sales everywhere in the U.S. I went through some old editions of North Platte's newspapers from the war years, and on most pages there were advertisements for famous brands—Waneita Schomer may have been a girl on a family farm with no electricity, but that didn't stop the merchandisers from trying to entice families like hers to be customers.

Montgomery Ward proclaimed that "in spite of rising costs, in spite of material shortages," it was cutting its prices on radios. "Here's your chance! Trade in your own set on this sensational 1942 Airline." The Airline was a stand-up console radio, bigger than most of today's television sets. Wards was selling it for "$59.88, less liberal trade-in." For families on budgets, a smaller radio could be had for $26.88. The war was the context of the sales pitch—"Now! Get Europe!"—and the little radio was promoted for its strong reception, necessary out on the plains: "Price cut on this powerful 7-tube! Look at the features—automatic tuning, tone control, loop aerial, big speaker! Includes rectifier!"

Burpee's seed company, operating out of headquarters at W. Atlee Burpee's Burpee Building in Philadelphia, took out ads offering the people of North Platte Burpee's Giant Zinnias in the "five best colors—scarlet, lavender, yellow, rose and white." Five packets of seeds could be purchased for a total price of ten cents. Pepsi-Cola, through its local bottling affiliate, was running a wartime promotion with North Platte bowling alley owners: "There's PEP in Pepsi-Cola to improve your bowling score! Insist on it always! Twice as big—twice as good!" The manufacturers of Camel cigarettes were positioning their product in an elite category, depicting in their advertising a man and woman on a ski holiday, enjoying Camels in the great outdoors; it was said to be "the cigarette of costlier tobaccos."

The nation and its famous products in the 1940s entered North Platte in myriad ways—from the Buicks on sale at Hahler Buick-Olds, to the name-brand couches, chairs and tables on display at Midwest Furniture Company on Dewey Street, where living room suites started at $49.50. So it was not that North Platte during the war was unaware of what the rest of America found to be alluring. North Platte knew—its citizens were being asked on a daily basis to purchase parts of that allure for themselves.

The affecting thing was not that North Platte knew about the East Coast and the big cities. The affecting thing was that men from those big cities were learning about North Platte. Men like George Dawson, of Manhattan.

"I was drafted into the Army Air Corps in January of 1943," said Mr. Dawson, who is now seventy-seven. "I had been a college student at City College of New York; before that I had gone to the High School of Commerce on Seventy-fourth Street.

"What they got when they took me into the service was a kid from Manhattan who had never been farther west than Hoboken. I became a B-24 navigator. There was a long trip across the country soon after I went into the service. A train trip to California."

On the train ride, he said, he began to see an America he had only heard or read about. "I liked the vistas," he said. "The immense spaces. I had not been used to that."

As the train approached North Platte, "I knew about the river, and Lewis and Clark and all that, from my studies in school. I did not expect what came next. A sergeant told us that we could get out of the train in North Platte, and we did—and here was this Union Pacific train station, loaded with tables. . . ."

As Paul Metro had when he had spoken with me from his hospital room, Mr. Dawson began to cry. He said:

"There was love there. . . ."

He had to stop.

"I'm sorry," he said after a few moments. "It wasn't Times Square, that depot. It wasn't Grand Central Station. But what that depot was . . . I was overwhelmed by the pure, simple generosity. We were treated as if we were their sons. They could not have treated their own sons with more kindness than they treated us."

He paused again. It was a while before he could continue.

"Nothing fazes a New York teenager," he said. "I took it in stride, that day. When you grow up in New York, New York is the city, anything else is the country. That day at the Canteen—all of this was happening to me with a bunch of guys. At some level when you're in your teens, you just accept the things that happen to you as the way things are.

"Nebraska was unknown territory to a New York kid. Before that day, I probably thought of Nebraska in terms of big, Scandinavian people. Probably what I saw in a Hollywood movie about Middle America. That day I learned something. I learned that the country was a hell of a lot bigger than Manhattan island. I found that out in North Platte."

He lives on Cape Cod now, he said: "above the state forest." He spent his life after the war as a sales representative for academic textbooks. He finds himself thinking of North Platte often, "with tender thoughts and feelings." There is a reason for that, he said:

"I don't know if you talk to many other fellows my age, but there is more looking back than looking forward. I had an uneven childhood. Things were not always the happiest in the house where I grew up. I needed a feeling of community more than most people, or so I suppose. That's what moved me so much in North Platte. It was their acceptance of me without question. I didn't grow up with that.

"Who were we, on that train? We were the hope of the world, at that moment. We were any kid on the street. We were all the same. We all wore the same uniform. We were 'our boys.' We were their boys."

He went into the service as a private, he said; he left as a second lieutenant. He went through North Platte twice more: "The next times I was there the feeling was the same, but I could let the guys on the train know what was waiting for them there."

He still recalls his first impression of the people who greeted him in the Canteen. "They seemed more secure, more centered than the people I was used to," he said. "I was not made to feel like an outsider. I brought that with me—the idea that I would be seen as different from them. But they were so welcoming."

He apologized again for being so emotional; he said he was surprising himself by the depth of his reaction. He said he had not talked about North Platte for a long time, and that he doubts he will ever get back there before he dies.

I asked him if there was anything he would like for me to say to the people of the town.

"Tell them they have a secret admirer on Cape Cod," he said.

"Tell them there's someone who loves them."

The horse was there every day, at Cornhusker Circle and McDonald. NO HORSES ALLOWED, the sign at the entrance of the nearby park admonished. But the horse was not inside the park—it just seemed to live in a city neighborhood, in a city where such a thing did not seem so unusual. In a city that even now did not feel like every other place in the world.

Thirteen

The twin sounds—I was getting accustomed to hearing them. One sound from the south of town, one from the north.

The sound from the south was the seamless roar of speeding cars. They were on Interstate 80, on their way farther west into Nebraska, or to Wyoming and Colorado. Even when you could not see the cars, you could hear them—their engines, their horns, their tires against the concrete. Through the trees, that sound never stopped, day or night. People moving, at the wheels of their own vehicles. No stoplights or stop signs on the interstate—never the noise of anyone or anything coming to a halt.

The sound from the north was only intermittent. If the constant whine of the automobile traffic never let up from the south of town, the whistle or rumble of a train coming from the north section, out past the viaducts, would catch you by surprise. The trains were freight trains, of course— no passenger train came anywhere near North Platte anymore. The freight tracks on the north side of the city were in the same place as had been the tracks that delivered the soldiers to the depot, all those years before. Maybe the people who now lived near the tracks kept a schedule in their heads—maybe they knew what time of day and night the different freight runs rolled through. But I could never make sense of it.

Every time I would hear the first hints of a train on its way, it surprised me anew. It was as if the cars to the south, on the interstate, were providing the solid bed of some song, the steady background part, and the trains up north were the veteran, temperamental lead singer, strolling into the studio to record his vocals only when he felt like it. The tune they made together was a pleasing one—I never tired of it. In North Platte, I began to hear it in my dreams.

I wasn't the only one.

"I was coming to North Platte with my father when I was three or four years old, to pick up groceries for his store," said Jim Beckius, seventy-four, who now lives there himself. "I was born in Stapleton, thirty miles north. My father had a grocery in Stapleton. He bought it in 1929, of all times to buy a business.

"It was called Beckius' Cash Grocery. But it seemed like it was all credit in those days. People just didn't have the money. There weren't many streets in that town, and the streets that there were didn't have names. There was just one real intersection, with street lamps on all four corners, and drinking fountains on two of the corners. As a boy I worked Saturdays and Sundays, stacking shelves and taking care of the eggs the farmers brought in to trade for groceries. Twelve or thirteen dollars was a huge order—that would buy enough groceries to fill up two or three boxes.

"My father helped out a lot of people in our town, by carrying them on credit so their families could eat. After he died, I looked at his records. There were people with twelve hundred or thirteen hundred dollars in bills, people who never had been able to pay him back.

"I was sixteen when I graduated from Baker Rural High School in 1943. There were only eighteen students in my graduating class. I went into the Navy as a combat air crewman. I had the glorified title of aviation radioman and gunner. I was afraid of the water. In the sandhills we didn't have any lakes, and I never did learn how to swim. I learned when I was in boot camp near Memphis. They throwed you in. I went right to the bottom, and they pulled me up. That's how they did it in the Navy during the war—if you couldn't swim, they gave you lessons. I got to be a pretty good swimmer, but I never did it after I got out of the Navy."

In December of 1943, he said, he was allowed to come home to Nebraska for a seven-day leave. "If you were from around North Platte, and you knew about the Canteen, you were kind of proud of it," he said. "You'd tell the other guys on the train: 'You're really going to be fed well, and it's not going to cost you one cent.' They didn't believe it. They said that no one ever got nothing for nothing.

"And then they got there, and saw a lot of ladies with a lot of food, and young girls of nineteen or twenty out on the platform with baskets of apples and magazines. . . . In pheasant season the people at the Canteen would fry up the pheasants for sandwiches, and some of these kids didn't know what a pheasant was. I'd have to tell them. I love pheasant—when I was a kid the hunting limit was twelve a day. I was seventeen that December when I came home for Christmas. My family didn't even know I was coming."

The depot that Christmas, he said, reminded him of when his father had taken him into North Platte all those times when he was so young. "I didn't know any of the people in North Platte when the train pulled in on my Christmas leave, but it felt like home," he said. "The next year, in December of 1944, I spent Christmas at Union

Station in Chicago. It was as empty as a tomb. I'd always heard about Union Station, how busy and full of people it was. But that Christmas there was no one there. It made me think about all the people at the depot in North Platte, all the people at the Canteen."

He served in the Pacific, got out of the Navy in August of 1946, and arrived back home on a bus. He got a job with the Union Pacific, at a time when rail travel was at a peak. "In October of 1947, there were thirty-two passenger trains a day stopping at the depot in North Platte. Think of what that means for a little town—thirty-two trains a day.

"The great trains—the *City of Los Angeles,* the *City of Denver,* the *City of San Francisco,* the *Challenger,* the *Gold Coast.* After the war, so many people were riding the trains. A lot of salesmen—even if they weren't staying in North Platte, they'd have time to run across the street and have a beer. They set foot in the town.

"It was better for the town, just seeing people all the time. And people from all over this part of Nebraska would come here to get *on* the trains. It makes a town feel like it has life. Once they made the highways better, and everyone started staying in their cars, the railroad just kind of started to fade away.

"It's been so long since we've had passenger trains. There would be no depot to go to, anyway. The town *sounds* different. You don't hear those old steam whistles. I hated the steam engines, at the time. You'd work on them, and they'd be hot in the summer and cold in the winter— you'd wish for something better. But when they left, I missed them. That sound, especially. Thirty-two trains a day—the sound was always in the air in this town. All the time—the sound was like the air itself."

It wasn't that the outside world had stopped coming to North Platte entirely. The world was simply delivered in different ways—more efficiently than by railroad trains.

The satellite dishes were a daily reminder of that to me. There was one that I kept passing—in front of a house on the 1500 block of Buffalo Bill Avenue. There was something about the juxtaposition of that—the snout of the dish aimed up toward the stars, ready to suck the world's images down to the ground, to a piece of road named for Buffalo Bill—that made me stop and pause next to the dish more than once.

But it had always been so. Only the technology had changed. During the war years, when first-run movies were America's primary form of visual entertainment, the theaters in North Platte showed the same films that audiences in New York and Chicago were seeing. During my time spent reading through North Platte newspapers from the 1940s, I found references to *Midnight Manhunt* starring William Gargan and Ann Savage, playing at the Paramount; *For Whom the Bell Tolls* starring Gary Cooper and Ingrid Bergman, playing at the Fox; *Man from Music Mountain* starring Roy Rogers and Trigger, playing at the State, all of the movie theaters downtown, near the Canteen.

And there had been radio: KODY in North Platte, WOW sending in its strong signal from Omaha, KFAB beaming in from Lincoln. The voices that floated out of the sky and into the homes here were voices that were being heard in Los Angeles, in Dallas, in Miami: *Clifton Utley Speaks* at 7:30 A.M., *The Fred Waring Show* at 10 A.M., *Young Dr. Malone* at 1 P.M., *Burns and Allen* at 6:30 P.M., *Elmer Davis News* at 7:55 P.M., *The Kay Kyser Program* at 9 P.M., *The Ramon Ramos Orchestra* at 11:30 P.M.

One day I spoke with a woman named Dorothy Townsend, who told me that the radio signals coming out of the sky weren't the only joy being delivered from above.

"It could be pretty bleak out in Nebraska during the war," said Mrs. Townsend, eighty-eight. "The towns we lived in were mostly very small, and there was not a lot for a person to do."

She and her husband had lived in Sutherland, she said, and one of the ways they had come up with to entertain themselves was to look at the stars.

"We would stop and watch the northern lights," she said. "We would pull off the road and sit and look at the sky. We would make an evening of it—we would come to North Platte to go to church, and then we would eat a bite before driving out to look at the stars."

She said that she remembers the beauty of it still:

"Bright, bright, shining lights. They looked like they came all the way from Alaska. In the winter, they seemed even brighter. My husband had a Ford—he was an electrical engineer, he helped make capacitors for Navy bombs—and he and I would sit there, not saying very much, just taking in the beauty of those stars above Nebraska."

She volunteered at the Canteen, she said, and in her spare time she would write letters to her two brothers. "They were my only two brothers, and they were in the service—my husband had four brothers in the service, so I would write to all six of those boys. You didn't send your letters to an address where they were actually fighting—you sent them to an APO address, and your letters were forwarded from there.

"I sent cookies, and they arrived all crumbled up. I would get letters back from the boys saying, 'We got your cookies—we ate the crumbles.'"

She laughed at the memory of that. "The meat we made for sandwiches at the Canteen wasn't crumbled, but it was ground up," she said. "When it was Sutherland's day at the Canteen, we would make our own meat and grind it up for sandwiches. We knew the boys on the trains would be in a hurry, so the meat sandwiches and the boiled eggs were ready for them when they came running in. People had told them on the train that if they wanted a certain kind of sandwich, they'd have to get there first. Of course, that wasn't true—we would give them any kind of sandwich they asked for.

"But they didn't know that, and they raced in—and sometimes they didn't know what to do. They just stood there. They didn't know what to say. So we would stand behind the counter and say, 'Would you like something?' They would nod yes, and we would say to them: 'Help yourself.'

"They were just new young boys, going to war."

Mrs. Townsend said that after the war she became a licensed ham-radio operator, to give herself a way to pass the hours. Pulling voices out of the sky, one at a time—not network radio broadcasts, but individual voices—pulling those voices into her Nebraska home.

"I would be talking to a person in Venezuela, and I would tell him where I was, and he would say, 'I went through North Platte during the war. I want you to know what it meant to me.' And then I would hear other voices coming on, from all over the world. 'North Platte? I've been to North Platte.' 'Did you say North Platte, ma'am? I once stopped in the Canteen.' I would tell them that I had worked at the Canteen, and then more and more voices would join in, saying that they had been there, and thanking me. Telling me how much it had meant to them, to have all of us waiting at the station for them."

There was one day and night in particular at the Canteen—an especially busy day and night for the troop trains—when seven thousand soldiers and sailors came into the depot. Seven thousand, in that one day. And they were all greeted, they were all fed, they were all thanked.

I asked Mrs. Townsend if the boys were aware that this had been the Sutherland ladies' day at the Canteen—if they realized that the women who were working so hard to make them happy were on hand because it was Sutherland's day to be there.

"No," she said softly. "That wasn't the point.

"The boys didn't know it was *our* day.

"It was *their* day."

Everything I was hearing about the town—every story, every remembrance—was told against the backdrop of the depot not being there anymore. Its absence from the city, for the people who had once worked inside the Canteen walls, was akin to a limb being missing from a person. Something was not whole, and never would be again.

When the trains had constantly brought visitors to North Platte—thirty-two times a day, even after the war was over, as Mr. Beckius had told me—the town, or so it seemed, had been something that it now was not. I wanted to find out what that truly meant, from those who had been around when the depot vanished—who had been in town when the sound of the passenger-train whistles stopped forever.

Fourteen

There were days when I discovered that entire lifetimes had been played out on the platform next to the tracks—it was as if the depot platform had been a theater stage. Home to dramas performed without the benefit of any script.

"I sold newspapers at the depot, when I was a boy," said Donald Land, sixty-nine, who now lives in Kansas City, Kansas. "We had lived in a town called Dickens, Nebraska, that had a population of about one hundred fifty. But work had kind of run out in Dickens for my father, so in the summer of 1942 he moved us to North Platte. I was ten."

He was the new boy at Cleveland Elementary School; his father had moved the family into a house on Roosevelt

Street, north of the Union Pacific tracks. Wanting something to do in the unfamiliar town, he decided to try to sell papers.

"I don't know what gave me the idea," he said. "I guess I just thought of it. I would go to the North Platte *Telegraph* office after school and pick up my papers. I paid for the papers, and sold them for a nickel apiece. I made two cents a paper.

"I would do it every day I could. I would wear just what I went to school in—a pair of Lee jeans. I would wear them, and a regular collared shirt, and go to the depot and wait for the troop trains to come in.

"The GIs would get off the trains and hurry toward the Canteen, and I'd have a handful of papers. I could sell a whole bundle and run back to get some more—the *Telegraph* office was only about a block and a half away. I'd hold the papers in my arm unfolded, and call out, 'Paper, paper!'"

That's what he was doing one day when he was startled to see a certain soldier getting off a troop train.

"It was my first cousin Edward Yonker, from Oberlin, Kansas," Mr. Land said. "My mother's sister's boy. I had no idea he was even in the service. He was about eighteen. This troop train pulled up, and I was calling 'Paper, paper!' as usual, and there was my cousin Edward, in an Army uniform.

"I probably hadn't seen him in five years—we used to

have family reunions, and I had last seen him at one of those. I just walked up to him with my papers in my hand, and said, 'Hi.' He was just as surprised to see me. I don't remember what our conversation was, except that he told me that seeing me made him homesick.

"He went into the Canteen and did whatever the soldiers did in there. I kind of waited for him on the platform. He came out with a sandwich, and I waved goodbye to him.

"The train pulled out. I remember thinking that I would have liked to go with him. When you're a young boy, you want to be a part of all that. I knew they were going off to war, but I guess I didn't really understand what that meant.

"When I got home that night, I told my mom. I said that I had seen Eddie."

And as for Eddie himself? What had happened to him?

Edward Yonker lives in Fort Collins, Colorado, and at seventy-six is that town's retired fire chief. Mr. Yonker told me that on the day he stopped in North Platte, "I was going across the country on my way to catch a ship." The ship was the USS *Shangri-La,* an aircraft carrier that would take him to the Pacific.

"I had just gotten out of high school in Kansas," Mr. Yonker said, "and I had signed up for the duration plus six months. I didn't really know what I was getting into—I was a kid from a dry-land farm.

"Our troop train was kind of open, and the black soot from the engine would come right into the cars. We were pretty black. I didn't expect to see anyone I knew at the North Platte station. My buddies knew I was from that part of the country, and they knew I was feeling lonely. One of the guys asked me, 'Are you going to go AWOL here?' I told him, 'I feel like it.'

"And then I walked from the train onto the platform, and I ran into Don, my little cousin, selling his newspapers. It made me feel so homesick—I was tempted to go home with him. The fact that the train station was so close to his parents' house . . .

"Instead I just went into the Canteen and got some sandwiches and milk. I went back onto the platform and asked Don about his family, and asked him if he had heard from my folks. I made the decision to get back on the train, of course. But I wished I could have stayed. I wished I could have stayed in Nebraska that day."

He continued on toward the war. But the platform next to the tracks was not finished with the Land family yet. There would be one more act in the play.

In 1951, Don—his paperboy days over—was in the service himself, stationed in England with the Air Force. The base was near the Mersey River. At the time, the Air Force base would send out invitations to local businesses, inviting young women who worked there to come to dances.

One young woman—really just a girl of sixteen, named

Jeanette Pattullo—was working as a secretary at the Edward Denton and Sons accounting firm. "I was going to night school also, to learn shorthand and typing," she told me. "My sister thought it would be fun to take me along. My employers scolded me for taking off from work early that day to get ready to take the GI bus to the dance."

That night, she met Donald Land. He was nineteen— only a few years removed from selling copies of the *Telegraph* outside the North Platte Canteen. "It might not have been love at first sight," she said. "But it was infatuation at first sight."

Before long they were married. Then Donald Land was sent back to the United States, to another Air Force base. Their baby son was born after he was already gone.

"So I set off for the United States on the liner *Britannic*," Mrs. Land told me. "Six days on the high seas for my baby and me. My uncle was the chief bedroom steward on the *Britannic*. I come from a family of seafaring people.

"We got to New York and I took a train all the way from New York City to Chicago. My mother-in-law and sister-in-law were going to meet me. The three of us, and the baby, were going to ride together to North Platte.

"I celebrated my eighteenth birthday in the Chicago train station. I didn't even realize it was my birthday—I was too busy taking care of the baby. Before I met Donald, I had never even heard of the state of Nebraska. It sounded

like something out of an American cowboy movie—
Texas, California, Arizona, Wyoming, all of those places.

"I can see all of it in my mind's eye now. The train, the
people. It was so hot that day I arrived in North Platte.
Donald was standing with the rest of his family on the
platform at the station as our train pulled in. It was June; I
hadn't seen him since February. He had never seen his son.

"I remember embracing him on the platform, in all the
heat. I was with my husband and our baby. The three of us
were together, starting our new life. Right on that platform at the train station."

They have been married for forty-eight years now. The
drama on the platform stage, at least for the family of
Donald Land, was an overwhelming success, with many
encores.

But the stage itself—the station platform by the
tracks—is long gone.

Around town, the verbal shorthand has it that the Union
Pacific passenger depot was torn down "in the middle of
the night."

The words are said with bite and bitterness. By 1973,
the passenger trains had been gone for two years. The
tracks still ran through North Platte, used for the Union
Pacific's freight operations. But the days of salesmen stepping off the trains and into the station, of visitors strolling

into downtown after a train trip to the city . . . those days were over, and they weren't coming back.

Still, for two years after 1971 the old depot remained. Some Union Pacific business offices operated out of the building, and it was lent out for community gatherings. It was a reminder—not just of the Canteen, but of what the city had felt like when it was a proud part of America's passenger rail system.

During the two years after the passenger trains disappeared, North Platte residents recall, there was talk from time to time that the depot might be demolished. No one really believed it; it was impossible to visualize downtown without its onetime lifesource.

So when, during the first week of November in 1973, the Union Pacific, without any announcement, moved in crews to knock the depot down, there was hurt in the town, and impotent outrage. The railroad, after it was done, explained that high-speed freight service was the company's source of revenue now, and that the old train station had no use in North Platte's post–passenger-train era. Some in the town were convinced that the railroad destroyed the depot *because* it feared passenger service might return—human passengers were a money loser for railroads by that part of the twentieth century, and anything a railroad company could do to discourage passenger traffic was considered fiscally smart.

At least that is what some of the town's longtime residents believed. But the Union Pacific could do whatever it wanted—it owned the depot and it owned the land. What the Union Pacific wanted was to get rid of the station. It happened quickly, and was done.

"They tore the depot down in the middle of the night," those who loved it say today. But there are few of those people left—and their numbers diminish daily, recorded on the obituary page of the *Telegraph*.

"You would hear the train whistle from off in the distance," said Lorene Huebner, seventy-six, "and it would send a tingle down your spine. It was a thrilling feeling that you didn't get anywhere else."

She was telling me about what went through her as she stood on that platform waiting for the soldiers to arrive. "The adrenaline would pump through you," she said. " 'Here comes the train.' I imagine the feeling is what it is like when you're skiing down a hill."

She was a teenager at the time; her family lived on a Nebraska farm near the town of Hershey, and her mother belonged to the Ladies Aid of the local Methodist church. "We teenage girls didn't get to go into town unless we had a mission," she said. "There were only eleven students in our class. So when Mom would go to the Canteen, I would want to go with her."

The Ladies Aid women would dress up on their Canteen day, Mrs. Huebner said, and so would she: "The women wore their Sunday clothes, and lot of them wore hats. They didn't have money to get their hair done at a beauty parlor, so they put the hats on.

"I was about sixteen. It was exciting to go to North Platte and see the handsome young sailor boys. Our parents didn't worry that we would run off into the wild blue with a boy. It was an impossibility. How would a girl run off on a troop train?"

It was the older girls from the Canteen who would tell the younger girls how much fun the place was: "They would tell us that it was overwhelming to meet all these guys in uniform. There's something about a man in uniform that demanded your esteem and made your heart thump. We were enthralled by them."

She said she liked to wear "a red dress with a keyhole neckline" to the Canteen: "I wore it because it was pretty flashy. Your heart got all aflutter when the train arrived. The ladies in the Canteen would tell you what they wanted you to do . . . hand out sandwiches, pass around apples in baskets. . . . There were no Styrofoam cups at that time, they just had the old china type at the Canteen, and each cup had to be washed and dried with a tea towel."

As indelible as the memories are now, she said, at the time "it was just part of your life. You didn't put an *X* on the woodwork—you just went to the Canteen when it was your day.

"That piano in the corner—some sailor boy would come in and start playing, and it was fabulous. It set the mood. I don't remember dancing—there was so much going on, it was pure bedlam."

Soon enough, though, "another train would be coming down the track. You'd say 'See you later, and good luck' to the boys getting back on one train, and you'd get ready for the next train. Sometimes a boy would tell one of the women that she reminded him of his mother, and he might get a hug and kiss from her before going."

At the end of each day spent at the Canteen, Mrs. Huebner said, "You would feel like you had done something worthwhile, for the glory of God and the glory of your nation. You would pray that those boys you had just seen would come back home. They were not much older than we were.

"And you would ask yourself if this was ever going to end."

She married "a farmer boy," and "we celebrated our fortieth anniversary. He died before our forty-first." For a while, she said, "I didn't care if my life went on or not." Now she is on her own, and she is surprised by some of the things that bother her.

"At ball games, you're supposed to stand with your hand over your heart for the national anthem," she said. "I go to high school basketball and football games, and I see people put their hands behind their backs, and I want to shout at them. Some don't take their hats off. Those darn ball caps, worn backward. During the anthem. I want to scream and shout to the high heavens. Show some respect!"

And of course she thinks of the Canteen, and of the long-ago boys in their uniforms:

"They were all young and scared, I would say. Lonesome. A lot of times they didn't know where they were going. We made them feel good for a few minutes, but probably by the time they were a few miles down the track, they had returned to lonesome and scared."

She said that what North Platte gave the soldiers was love. Not the trivial type, either.

"There are different kinds of love," she said. "You can love a strawberry shortcake. You can love a ride to town.

"But this . . . this was real. *This* was love."

Every time I would stop by the rectangle of land where the train station used to be, I would find only vagrants there. It seemed to be a gathering place for men with nowhere else to go; I would see them drinking liquor from bottles concealed in brown paper bags, or sometimes not concealed at all.

Day and night, they would sit with their backs to Front Street, staring toward the tracks. I thought about what might have happened to the old depot, had the Union

Pacific let it stay up. Without a reason for being there, without railroad employees and passengers inside the building full-time, it might have turned into something tawdry: a drug house, a place for squatters, somewhere dangerous. A blight on the town.

So its absence probably made some sense. But if the old depot—the old Canteen—was absent in a physical sense, it was present in other ways, ways that might have been just as important. I got one more reminder of that when I met the mayor.

Jim Whitaker, the mayor of North Platte, was carrying a thank-you note around town. It had just arrived—almost sixty years after the fact.

The note was from Bill Dye, a survivor from the aircraft carrier USS *Lexington,* which was sunk in the Battle of the Coral Sea on May 8, 1942. Mr. Dye, who had mailed the note from his home in Estes Park, Colorado, said he was sending the belated thanks on behalf of all the survivors from his carrier who had been treated so kindly at the Canteen.

"All these years later," Mayor Whitaker told me. "Isn't that something—that they would write to thank us all these years later?"

I called Bill Dye, who is seventy-nine. "Some of us

were talking, and we realized that none of us will ever forget the lovely, lovely way we were greeted in that place," he said. "Especially after what we had been through."

The *Lexington,* he said, was hit by two torpedoes and three bombs. "It was hopeless to save the ship," he said. "I was in the water an hour. I was picked up by a destroyer—at nineteen you think you're going to live forever, and you thought the *Lexington* could do anything. I had only my dungarees and my shirt. I think every man in the water left his shoes on the flight deck. The explosions, and the smoke . . . and then the ship was gone."

It took three weeks for him to get from the Coral Sea to San Diego—he was transferred from a destroyer to a cruiser to a troop ship. Then he was sent to fight again in North Africa. And at some point during all of this, while being shuttled across the United States, his train pulled into North Platte.

"You have to understand," he said, "on that train, you had no bunk. You sat up for three days. You had no shower. You were pretty weary.

"And then . . . you find this unexpected bouquet of nice people."

Along with his note to the mayor, Mr. Dye had sent a giant thank-you card, signed by fifty of his fellow crewmen from the *Lexington*—fifty men who were survivors still. Late in their lives, they wanted to make sure that while they were still on this earth they could find a way to express their gratitude to a place that had embraced them when they were young and very much in need of it.

"Those people in that town," Mr. Dye told me, "they were solid rocks. Whoever figured out how to start that Canteen and run that Canteen should be president of the United States."

No president had been to North Platte in quite some time.

But a governor had once endeavored to make one building in town as elegant as anything in Paris or London—as I was, with some melancholy, about to find out.

Fifteen

"Downtown at the time was tremendously active, day and night—and this grand hotel, the Pawnee Hotel, was the center of everything."

Larry McWilliams—the man who, with his young buddies when he was a boy, used to watch North Platte's painted women of the evening come in for cherry Cokes at the drugstore—was filling me in on the Pawnee. The hotel had opened in 1929 with ambitions of becoming the diamond of the Nebraska prairies. The man behind it had been North Platte's most successful politician of that era— M. Keith Neville, who was born in town and had risen to become governor of Nebraska in 1917.

After being defeated for a second term, Governor Neville came home, with the idea of building a hotel of great luxury and the highest standards. And he did it—on Fifth Street he constructed the Pawnee, intending to make it so beautiful that people's eyes would widen when they gazed upon it.

"It was an elite, elegant hotel," Larry McWilliams told me. "It was the absolute nucleus of North Platte society. The Crystal Ballroom was beyond compare, at least in this part of the country. The high school proms were held there, the big annual festivals, the May Ball, weddings, retirements . . . any social occasion that aspired to be noticed took place in the Crystal Ballroom of the Pawnee.

"Governor Neville also built the Fox Theater, across the street from the Pawnee—a beautiful theater for live entertainment. He really wanted North Platte to be something."

Few of the soldiers who arrived at the Canteen, just a few blocks away, ever got the chance to visit the Pawnee— they only had those ten minutes before the trains rolled out again. Which seems a shame—the Pawnee, even before the war, endeavored to be a signal from the city to the sophisticated outside world: We will try to meet you on your own terms. We will do our best, even if most of America has no idea we are here.

And today? The Pawnee today?

"It's sort of a retirement home," Larry McWilliams said. "Nothing much really happens there."

Sad to say, he was being too kind.

I walked up to the Pawnee in midafternoon. "Retirement home" appeared to be a euphemism. One step into the main foyer, and I could tell that the Pawnee was an end-of-the-line place—it had the feel of an inner-city mission, a shelter, for those low on both income and hope. The vacant stares on the faces of some of the men and women who sat motionless in the heat, the one fellow talking rapidly to himself with no one in earshot to listen, the utter torpor . . .

It would have been a sorrowful enough scene in any setting, but I had come to the Pawnee with foreknowledge of what the hotel had been. Not that those present on this day weren't trying; a man in the main lobby attempted to hold a religious service for those whose attention he could attract, and the ladies behind the front desk were friendly and helpful, although seemingly surprised to have a visitor who didn't have a relative in residence. They told me to feel free to have a look around.

It was difficult, standing in the middle of all this, to envision it as it once had been—the glittering, glowing beacon of the town, sending out its enchanting invitation to the residents of all of the counties in every direction, the place in this part of Nebraska to inspire deep-in-the-night dreams and summon up glorious memories.

The air on the main floor was heavy, the fragrance

unpleasant, and as I headed for a stairway that would take me to the Crystal Ballroom, I thought about a conversation I'd had with one of the old Canteen volunteers, a woman who had more than a passing familiarity with the Pawnee. She was Governor Neville's daughter—and she had lived within these walls, long ago.

"My dad was known as 'the boy governor,' " said Virginia Neville Robertson, eighty-eight. "He was only thirty-two when be became governor of Nebraska. And when he came home, he wanted to give the town something great, and what he gave the town was the Pawnee."

The hotel could be seen for miles around, she said: "It was the only eight-story building in town. Those were the days when a great hotel was the central point in any city, and the Pawnee was North Platte's great hotel. But to me, it was a home. We lived in the hotel, and raised our children in the hotel."

She met the man who would become her husband, Donald Robertson, in 1934 when they were students at the University of Nebraska in Lincoln; they were married three years later. "He became a desk clerk at the Orrington Hotel in Evanston, Illinois," she said, "and then he was a desk clerk at the Park Lane in Toledo, Ohio, the city where both of our daughters were born.

"But then the war started, and he volunteered. I came

home to North Platte. My father was such a great believer in the town. So while my husband was off at war, I helped out at the Canteen. I was married, so I wouldn't really have fit in as a platform girl—I mostly stood behind the counter with my mother and handed out cake.

"My father would come down to the Canteen quite a bit. I doubt that many of the soldiers who came in knew that he was the former governor of Nebraska. He was a very, very quiet man—he always used to tell us four girls, 'Don't ever blow your own horn.' "

When the war ended, she said, her husband joined her in North Platte, and became the manager of the Pawnee. "We lived there from 1945 until the hotel was sold in 1973," she said. "Especially with children, to live in a hotel was unusual, but we liked it. The Tom-Tom Room coffee shop off the lobby was sort of the town's gathering place, and because my parents lived in North Platte, and my three sisters, we could get together as a family at the hotel, and the children could play after school."

She was proud of her father, she said—both as a dad, and as a man who wanted to do right by his hometown. "For so many years after he died, there was nothing in North Platte that bore his name," she told me. "But now the old Fox Theater, across the street from the Pawnee, is used for plays put on by the North Platte Community Playhouse, and the building is called the Neville Center for the Performing Arts. So finally his name is on a building, which I think would have pleased him, although he never would have said it."

Mrs. Robertson's husband died in 1978, five years after they moved out of the Pawnee. "I look back with great fondness on our years living in the hotel," she said. "It was the finest, most beautiful place that North Platte ever had."

The walls were peeling in the Crystal Ballroom. Water damage had left ugly streaks.

Where once the citizens of North Platte had celebrated their most consequential occasions by dancing late into the night and dining decorously from china plates with silver settings, cafeteria-style tables with rough wooden surfaces had been set up in the ballroom. Yet someone had taken care to try to make things homelike for the current residents of the Pawnee: A little vase with fresh flowers sat atop each table.

The crystal was still there, if you looked up: Six chandeliers, in varying states of repair, descended from the ceiling. A rostrum—this must have been the orchestra stand—abutted the wall closest to the exterior hallway.

I walked around the second floor of the hotel, trying to imagine what it must have been like to come here when doormen greeted arriving guests, and parking attendants whisked automobiles away, and salesmen working their Western routes stopped for the night at the Pawnee after having alighted from their trains. What it must have been like for the ranchers and farmers from the surrounding towns to come here for the weddings of their sons and daughters, giving their children a ceremonious send-off on what the parents hoped would be a serene and successful life.

There was a framed newspaper front page on the wall of a corridor near the ballroom; it was from the *Telegraph* of October 16, 1929. The banner headline announced HOTEL MAKES FORMAL BOW TO PUBLIC TONIGHT, the pride and optimism for the town almost palpable in the week just before, unbeknownst to anyone—the hotel's owners, the citizens of North Platte, the person who wrote the newspaper story—the United States would be plunged into the most terrible economic depression in its history.

And then, later, would come the war. I took the stairs back down to the first floor, and noticed a squared-off sign for what had once been the Tom-Tom Room, the popular and busy coffee shop the governor's daughter had told me about. The fellow in the lobby who had been trying to recruit congregants for the day's religious service was still at it. He didn't seem to be getting many takers. A man and a woman, perspiring profusely in the heat, saying nothing, worked a jigsaw puzzle, one helping the other.

"When I arrived in North Platte, I was feeling about as low as a guy can feel."

Paul Gardner, seventy-nine, who now lives in Scottsdale, Arizona, was describing to me what had happened just before he got on the train that would stop at the Canteen—what it was that had so dispirited him.

"I didn't want to go home," he said. "But they wanted to discharge me."

He was a corporal in the Army Air Corps, which he thought was leading him toward something he had longed for his whole life. "I wanted to fly since I was ten years old," he said. "That's the year they were opening a new airfield near where I lived, in Lima, Ohio. I walked five miles to see the bucket-scrapers drawn by horses—ten years old, and I would walk to watch them build that airport."

He enlisted in 1942, right out of Lima South High School, and was in Army Air Corps training in California when he went up on a day he now knows he shouldn't have. He had a cold. Nothing more serious than that.

"I got bilateral infections," he said. "Both ears—infections of the middle ears. I was deaf for five days, and I was hospitalized. I shouldn't have flown with that cold—I got my hearing back, but the infections had thrown off my sense of balance. Instead of graduating from pilot training, I was in the hospital taking sulfa drugs.

"The medical board washed me out. You either achieve a goal or you don't—my goal was to fly, to do whatever I could as a pilot to help the war effort, and the medical board told me it wasn't going to happen. I was twenty-two or something like that. I cried.

"It's not like a baseball player who gets a year or two to recover, and goes on with his career. To fly in the war, they wanted you in training, and they wanted your training to be on schedule. Either you could do it or you couldn't. My balance was all screwed up. They told me they couldn't put me up there in the air in a position of responsibility. They told me my health would prevent me from carrying it out.

"We had to win the war. So many things depended on us. . . ."

They told him to go home to Lima for two weeks, to see his family. "I had to go north to Sacramento to get on a passenger train," he said. "I was on sick leave. There were about fifty or sixty of us servicemen on the passenger train, and as we were riding across Nebraska we were told that in the town coming up, they had a Canteen for servicemen.

"So that was my introduction to North Platte. My goodness . . . here I was, feeling the lowest of my life, feeling like a complete failure, and we got off the train and those people made me feel pretty special. Just the feeling of being *appreciated*. They were so *nice* to me. They didn't know me; they didn't know anything about me. But I was in the uniform of my country, and that was good enough for them."

He ended up spending his adulthood working for Westinghouse as a quality control specialist; his health has not been good in recent years: "I've got one of the worst hearts in the country." He knows that by being disqualified from flying in the war, he may have been given a chance at life that other young men did not have.

"There was such a feeling among young men," he said. "It was a feeling of 'I don't know what tomorrow will bring.' You got a letter from home—'so-and-so died in Europe.' You lost classmates from high school, from your hometown. It only has to happen to you once to stick in your mind. One of my football teammates at Lima South High School—Bob Brodbeck. I was the second-string quarterback, he was an end. I never got to say goodbye to him. I learned that he was killed in action in a letter from my mother."

So by being grounded by the Army Air Corps medical board, Paul Gardner may have been spared a similar fate. He didn't fly in the war—and he didn't die in the war. It still bothers him, that he was not allowed to become a combat aviator.

And it still is with him, that memory of the brief time in North Platte that meant so much to him.

"There are moments in a young man's life when it means a great deal if someone seems to appreciate him," he said. "You have no idea what the respect I was shown by those people at the Canteen did for me. I was feeling like I was no good to anybody. And those people made me feel that I mattered."

I left the Pawnee Hotel, taking one look back at it from the corner of Fifth and Bailey, trying to imagine the pride felt by former Governor Neville—and the pride felt by the town—on that night in 1929 when it had opened its doors.

So ornate, so aristocratic . . . it had been meant as a monument, a gorgeous monument inside of which people could sleep. It had been meant not only as the boy governor's gift to the town . . . but in a way, undoubtedly, as his gift to himself. His try for immortality.

There weren't many buildings like that going up anymore, at least not in North Platte. I had seen some obviously expensive homes during my walks through town—some of them of recent vintage. One in particular, a sprawling place with a fence around it, the kind of home you would find in the most exclusive suburbs of America's biggest cities—every time I passed it, I tried to guess who lived inside.

The biggest industrialist in town, whoever that might be? Another governor or senator, out of office and having cashed in on his connections? The leading banker, or the head of the hospital? In the meritocracy of this part of Nebraska, who had risen to the top—who had constructed for himself this stately mansion?

It turned out not to be a he, but a she. And as for the meritocracy . . . as for the idea that the person who built the large and costly house did so after working her way to the top of the world of commerce or politics, the way Governor Neville had before erecting the Pawnee on the plains . . .

This was a new era—all over the United States, and right here.

The owner of the spectacular home—the master of the manor?

She had been an employee of the North Platte post office.

She had won $50 million in the Powerball lottery.

She lived about as far from the tracks as a person could live.

Sixteen

I was beginning to understand just how the sandhills—the defining physical characteristic of this part of Nebraska, the boundless dry sea of rolling dunes that made everything feel the way it felt—had shaped the lives of the men and women of the Canteen days. How the sandhills had made everything in the world of those men and women seem vast beyond comprehending, yet paradoxically small and confined, all at the same time.

I spoke with people who had grown up surrounded by all of this—Marian and Kent Peterson were two of those people—and as I listened to them, I could almost see them here in the 1940s, see them as they would excitedly come into town with their parents to meet the troop trains.

Marian Peterson—she was Marian Rodine then—is sixty-eight. The way she recalls being a young girl in the sandhills, "We went to church, or grocery shopping, but we didn't go to the movies. I grew up near Gothenburg, and I don't think I saw a movie until I was in high school."

Which is why the Canteen was such a thrill for her, at the age of eight or nine. "I would go with my mother's ladies' group—it was a church group called the New Hope Dorcas. My mother never learned to drive, so my dad did the driving to the Canteen. He drove, and carried the food.

"It was such a different way of life for me to see. The sailor boys were so good-looking, to a young girl like me. The sailors and the soldiers would just come rushing in the door, like they couldn't believe there was enough food for them. They would play the piano and sing . . . that was so *interesting* for me, never having seen anything or done anything like that."

Her knowledge of the war, she said, mostly came from listening to family conversations. "I don't know if we had a newspaper in the house or not," she said. "We mostly gathered around the radio not to listen to war news, but to soap operas. *Fibber McGee and Molly.* We'd sit there and listen to it together, and play dominoes.

"The war, to us, was not something that came out of the radio. It was the boys at the Canteen."

In a different part of the sandhills, the man she would

marry—Kent Peterson, now sixty-seven—was growing up, too. They had yet to meet.

"My family lived near Brady, between Gothenburg and North Platte," he said. "We had a pretty modern radio, for the time—my uncle had a hardware store, and we came into town from the farm and bought one there. We kept our radio in the kitchen—that's where you gathered, that's where you had your heating stove, and your food.

"One of our neighbors, Mr. Craig, would come over and listen to the radio with us. He liked the fights. So if boxing was going on, we'd have to listen to boxing. War would have to wait.

"You would listen to the radio with your family in the kitchen, and you just warmed up real good and jumped in bed. You would heat up a flatiron, wrap a blanket around it, and put it in your bed down by your feet. That would keep you warm for most of the night."

He said he would go to the Canteen with the group of volunteers from Brady, often in the evening: "We'd get out of school, and hurry up and milk the cows. I definitely liked going to the Canteen—we couldn't wait to get those fourteen cows milked so we could get started.

"The servicemen from the trains were always so neatly dressed. It was such a rat race, with them hurrying around—I didn't want to be like those guys. It was kind of scary. I was a little boy from a farm, nine years old. It was too much for me to want to be a part of.

"But it was fascinating to see. My family and I always just wore overalls all the time—I had never seen anyone with uniforms like that. It was such an event, riding the twenty-eight miles to see all that at the Canteen.

"Muddy roads, and no lights along the roads at night—but I hadn't seen anything in the world. I might have gotten to town once a month, if that. To take cream in to sell to the grocer. We'd get seven dollars for a five-gallon can of cream, and we'd spend six dollars of that on groceries, and save the other dollar for church."

That was sandhills life, for a young person as the war was beginning. Eventually, he and Marian Rodine would meet.

"I was a cheerleader at Gothenburg High School," she said. "He tells me he would come from Brady to the football games and see me cheer. I graduated from high school in 1950, and I didn't know him. I went away to the Bonnell Beauty Academy in Hastings, and then I came back to Gothenburg to work at the La Grace Beauty Shop. I met Kent at church, at an evening Christmas program. He escorted me back to my home that night. We got married in 1954."

Kent Peterson said: "The cheerleader girls were pretty good-looking, and she was the prettiest one of the bunch. I'd go down to the fence and pretend to be watching the game. You had to be on the move, at that fence—you couldn't just stand there. But the reason I was there was to look at her."

I asked him if, when they finally met, they had fallen in love right away.

"Oh, I don't know," he said, half a century later, laughter in his voice. "We're still working on it."

I rode north on Route 83, past the viaduct, past the tracks, past the old Elks Club and the softball complex, out of town. I wanted to see the sandhills at their most unpopulated.

At one time—as late as the 1860s—this was known as the Great American Desert; some topographical experts of that era declared that it was unfit for human habitation. They were wrong; although the sandhills are, in essence, literal dunes, their fragile surface is held in place by the tenacious roots of native grasses. The grass stabilizes the hills; as long as the grass is safeguarded and kept intact, the sandhills provide the best grazing land for cattle on the North American continent, and a stunning portrait of glorious isolation for those who live here—and especially for those who are seeing all of this for the first time.

As I proceeded farther out of town—in the direction of Stapleton, and Thedford, and Lewanna and Brownlee and Valentine, all the villages leading to the bottom border of South Dakota—the houses became fewer, and the stretches of land between the farms grew longer and longer. There was green all around, in every direction; contrary to the name, the sandhills in summer had the lush and spellbinding look of a perfect and undulating putting green, the largest putting green imaginable at the most far-flung and expansive golf course in the world.

There were moments—almost inexpressibly splendorous moments—when the sandhills were all that seemed to exist. Nothing but them, as far as the eye could behold. I thought about Mr. and Mrs. Peterson, and what it must have been like when they were very young, coming with their parents into North Platte, which, surrounded by the sandhills, must have seemed like Manhattan. And I thought, as I was always thinking now, of those soldiers on the trains. Of what they must have pondered as they passed through this country.

"Fourteen months ago I lost my wife of forty-eight years," said Lawrence W. Jones, seventy-seven, who lives in Nacogdoches, Texas. "I'm coming out of a tunnel. I was just numb for seven or eight months."

He was explaining that he felt all right—or at least felt good enough to talk with me about the Canteen, and why it has such an enduring place in his heart.

"It would have been about the first part of August in 1943," he said. "I was a tailgunner in the Army Air Corps. I wasn't even old enough to buy a drink.

"I had volunteered out of high school in Sarasota, Florida. There was nothing to do in that town—the only jobs available were working at soda fountains, at least when the circus left its winter headquarters and was on the road.

"I had graduated from a six-week course in gunnery school at Buckingham Air Force Base in Florida, and they sent us west on a train, to Salt Lake City. There were probably five hundred military people on that train—field artillery, infantry, air corps. And after that long trip across the country, at five-thirty one afternoon we pulled into this place none of us knew anything about.

"We looked out the windows, and there were these women talking to us, passing us sandwiches and everything. They said, 'Are you going to get off the train?' We said, 'We don't know if we're allowed.' They said, 'We've got it fixed—you can get off the train.'"

And so, he said, they all did. "We lived in our uniforms—we never wore civilian clothes. We were all in our uniforms when we walked into the train station . . . and there were these plank tables, loaded down with every kind of food you could imagine. Homemade cakes, pies, sandwiches, Coca-Cola . . . We could not get over it. Out in the middle of nowhere, or at least to us it was the middle of nowhere, and it was getting toward dusk . . . this was like a miracle.

"I got to thinking in my own mind: 'Where do all these people come from?' So I asked. And they told me that they were from all over that part of Nebraska, some from a hundred miles away. The farmers' wives made chicken for us, and they brought milk from the farms. . . .

"And they did it day after day after day after day. We were there for so few minutes, and then it was 'All right, load up the cookies, get back on the train.' Puff, puff, and we were gone."

As the train pulled out of North Platte, he said, he couldn't stop thinking about what he had just seen. "Those people spent all that time and donated all that money—to get the sugar and all that stuff. They gave up their own ration stamps. They were using their ration stamps for us. We all knew what that meant. I wrote home about it."

He was a tailgunner in B-24 Liberators over Europe and North Africa. "Wars are not about killing, and wars are not about dying," he said. "Wars are about love. That's what you remember. All the other stuff is incidental. The people you depended on, though, the people you came to love . . .

"It's the bulletin board syndrome. You got up in the morning and you went to the bulletin board to see if your name was on it. Whether you would be flying a mission. We were good at our jobs. We stayed good to stay alive.

"Flying those missions, everything was in slow motion. You're twenty-five thousand feet above the ground, and you have no feeling of movement. You watch the contrails behind the ship . . . the heat, mixed with the cold moisture in the air, behind each engine . . . The only noise you would hear is the drone of those engines. I'm all by myself back there, seeing that. I felt like I flew the whole war by myself, back there in the tail alone."

As he began to talk about North Platte again, I heard something in his voice. It was the sound I had come to recognize in these men.

"I don't know," he said, crying. "It's something that comes over you every once in a while. We had heard that there were women in that Canteen who had lost their sons. And they would come down there. To see us.

"I think that we really were all in it together. The women who were there—their personal lives, and the lives of our own mothers . . .

"It was either going to be freedom or slavery. If we had lost the war, the men would have been in slave camps, and the women, if they were good-looking, would have been in the officers' clubs of the enemy who defeated us, doing their 'duty.' Our enemies would have confiscated all of our wealth, all of our nation's art . . . everything. We never think about that, because we won the war. But if we had lost . . .

"You think back to the war, and it's not the shootings and bombings you think about. It's the relationships with people, and some of them, you realize now, you hardly even knew, but they still meant so much to you. When veterans get together, they talk about this mission and that mission. But what they really mean is: 'What happened to so-and-so?'

"That's how I feel about North Platte. What happened to all of those people? They were like our mothers and our sisters. How did they know to do that for us? How did they know how much it would mean?"

On Route 83, with the sandhills promising to roll on forever, I decided it was time to head south again—to turn around and go back to town.

That shoreless sea of green, no matter which way I looked—I tried to picture it in winter, covered in snow and ice, or in autumn when the grass turned brittle. The people who had driven through here every month of the year so they could volunteer at the Canteen—they must have felt that they were hurrying toward something: hurrying toward a world they had created at the depot, a world where the distant march of those war years had become authentic and present and a daily part of their very fabric. Hurrying through these silent miles toward something great.

The towering grain elevators just north of the Union Pacific tracks told me that I was approaching the city itself. Painted across the tops of the elevators, linking them, were the two words, as definitional as an urban skyline, an unflowery and unambiguous welcome: NORTH PLATTE. And painted beneath those words, as if to patiently explain that the grain elevators were for work, not for picture-postcard show: A FARMER-RANCHER OWNED SERVICE.

Underneath them, if you knew where to look, was the place where a train station once stood.

"They were oil roads," said Helen Johnson, seventy-three. "That's how we got to North Platte—on dirt roads that had oil on the surface to keep down the dust."

She grew up in Brule, a town of 410, sixty miles west of the Canteen. "Brule and Big Springs had their Canteen days together, because the communities were so small," she said. "My folks lived on a farm, so we had plenty of eggs and meat. My mom made meat sandwiches for the boys. It was easy to convince me to go. There was food in my dad's car from top to bottom. The trunk was chock-full of sandwiches. We would hold the cakes as we rode."

She was just thirteen when she started going. "You were delighted to greet the guys and give them home cooking, and as a teenager you liked to see the boys. But you also had this feeling in your heart that some would not make it home. You would look at them and feel a lump in your stomach thinking about their future."

Although there was never an official announcement that a train full of soldiers was coming, "It seems that we sort of knew ten or fifteen minutes before. The boys were anxious as they got off the train—I suppose there was a kind of wonderment in their faces. But, oh, when they got inside and saw the food laid out for them—I'm sure some of them hadn't had that kind of cooking for a while."

Seeing the young soldiers, "It made me realize all the more that the war was a serious thing. We had a radio on our farm, but the battery would always go down, and we would miss the end of the news reports about the war. We didn't have electricity on the farm—we had our own batteries, and a gas generator. The radio was in the living room—it was a console. I have to smile thinking about it, because it always happened—just when you thought you were going to listen to a whole program, the battery would go."

So her main contact with the war, the actuality of the war for her, was the time spent with the young men at the Canteen. "It all happened in such a flurry, each train," she said. "You would very quickly start putting out your sandwiches on plates, and pouring drinks so they would be ready. All of us would man our posts. The boys had such a fleeting time with us.

"There just wasn't time to get to know them. The faces all became a blur by the end of the day. But they were all real to us—and I think they were thankful that for a few minutes maybe they didn't have to think about the war."

After the last train of the day had pulled out, she said, "Of course the Canteen seemed very, very silent and vacant. The ride back to our farm in Brule would be quiet and pretty somber. All the food containers would be empty.

"You never really wiped away the thought of the boys. You could still see them climbing off the trains, and then filing back on. They knew not how many days they had left, or where they were going. They looked so young— they *were* so young—but they never said anything about it. It all went so fast at the Canteen, and they knew the train was pulling out shortly.

"At night on the farm, it would pass through your mind. The railroad tracks in our part of Nebraska were not even a quarter mile from our farmhouse, and every time you would hear a train, you would wonder if it was a troop train. If it was some of our boys."

The world has changed in many ways since those days and nights, but the sandhills have remained constant. And Mrs. Johnson, who still lives there, has never forgotten what she would do, as a teenage girl, whenever the young soldiers hurried from the Canteen and back onto their trains.

"I would pray," she said. "For all of them. I would watch them get onto the train, and I would ask the Lord to bless and keep them. I wanted to keep smiling, in case they turned around to look at us as they left. But I was praying for them, with my eyes open."

Seventeen

There was an airport in town, although I had not seen it in all my meanderings through North Platte, and had not even heard anyone mention it.

But I knew it existed—and that it was not just a strip for private planes, but a field that handled commercial aircraft. In most American cities that once depended on the railroads for interstate passenger service, the local airports were the reason train travel had dwindled. But if that was the case here—if the old Union Pacific depot had been torn down in large part because the North Platte airport had put it out of business—then the airport must have earned its success very quietly. Either that, or something else was going on.

On the most sweltering afternoon of my time in town, I decided to go out and take a look. If nothing else, maybe the place was air-conditioned.

America's airports, in coming to dominance, had shrunk the country—had made great distances seem trifling. Part of that was illusion, but the concept of the outside world itself had without question been altered with the coming of the jets. If you could be anywhere in the continental United States within a few hours, how vast or detached could the world outside your town's borders possibly be?

I thought about that as I spoke with Marjorie Pinkerton, seventy-two, who had helped out at the Canteen as a young teenager when her family had lived on a farm near Shelton. She would ride a train to North Platte to volunteer along with her older sister. "I felt like just a tagalong," she said. "Being the little sister, I just went wherever she went."

What Mrs. Pinkerton told me about her knowledge of the war, though—about how, as a young girl, she had kept current on the events overseas—made the new American dismissiveness of time and distance seem like the profound and pervading change that it has been. Because she learned her war news at a movie theater.

"That's one of the reasons that I don't think of the trips to the Canteen as being exactly *fun*," she said. "It was kind of a scary thing, because of what I was thinking about the soldiers.

"When we would go to the Saturday night movies in Shelton, they would always have a newsreel. We had a little theater in town—the Conroy family, right in Shelton, owned the theater, and it had just one aisle, and no balcony. The girl who would take us to our seats—she was Kathleen Moog, she was my best friend, we were in the same grade in school—would have a flashlight to show us where to sit. Bill Conroy—he was in my class, also—would run the movie projector.

"We would wait for Bill to make the movie start—we would be ready to see *Mrs. Miniver*, or whatever the movie was that night. But first there would be the newsreel. Without television back then, that is how we knew what the war looked like. The battle coverage.

"It was black-and-white, and it would really bring it home. When you just heard about the war, you had to imagine things and picture it in your mind. But the newsreels made you see it. When troop trains would roll through Shelton, we would wave at them, and the boys would wave right back at us, out the windows of the train. Right there next to Main Street, across from the movie theater—troop trains, right through Shelton. Whenever that would happen, I couldn't look at the boys in the train windows without thinking of the battles in the newsreels."

The local feel of much of America—the feel that lasted right up until air travel became commonplace—even played a role in how she would meet her husband. Their meeting was a byproduct of a local evening newspaper—sort of.

"In 1950 I had moved to North Platte, and I was a teacher at the Franklin Grade School," Mrs. Pinkerton said. "I taught third grade. I was walking home after school one day, and I saw a crippled lady—she was on crutches—on Third Street. She was in front of her house, and the newspaper was on the grass. I suppose the paperboy hadn't thrown it far enough.

"She was on those crutches, so I picked the paper up and carried it to her. We talked, and I still had a few blocks to walk to where I lived, and she said to me, 'Well, come over this weekend, and we'll have lunch together.' So that Saturday I went to her house, and we started visiting back and forth, and she told me that she had a grandson. She took me to her china closet and said, 'This is his picture.'

"It was his high school graduation picture, and I thought to myself, 'Gee, I don't want to meet him.' She told me that he lived with his mom and dad on a ranch in the sandhills, about thirty miles away. His graduation picture didn't do anything for me, but she kept insisting and kept insisting. So I said, 'All right, I'll meet him.' And she had him come to the house, and we did meet."

He was Harry Pinkerton, and they were married for forty-six years, until his death in 1999. They had two children. "It was a perfect match," Mrs. Pinkerton said.

And it began on a side street with a mistossed copy of the evening newspaper as she walked home from work late one afternoon. It couldn't happen that way now. There is no longer an evening paper in North Platte.

Out Fourth Street, past downtown to the east, onto the old Lincoln Highway and through fields and countryside, I looked both to my left and to my right, trying to find the airport.

I almost missed it. An unprepossessing building miles out of the main city, it had a free parking lot with no attendants—the kind and size of parking lot you might find at a suburban branch bank. There were no cars parked next to the terminal building; anyone could pull right up to the door and walk in.

No passengers, either—that's what I discovered as soon as I got inside. At the single boarding gate of the airport—the place is called Lee Bird Field, named to honor a North Platte family's son who was killed while training as a military pilot in 1918—no agents were visible. I checked the schedule—there were only two flights today, and both had already departed.

I approached the ticket counter; a woman told me she was unable to issue tickets to passengers, because an agreement between United Airlines and Great Lakes Airlines to run a joint operation had recently fallen apart, and until the details of Great Lakes' unshared proprietorship of the route could be worked out, purchases had to be made through a 1-800 number, and not at the airport.

Not that this affected all that many passengers; on a weekend day like this one, there were just the two flights to Denver (on Beechcraft twin-engine propeller planes); on weekdays there were three flights. The airplanes held only nineteen passengers, and were seldom full.

This is what had helped to do away with the once-bustling railroad station downtown; this, at one time, had seemed to be the lustrous future of long-distance transportation out of and into North Platte. Frontier Airlines had for a time flown jets—737s—into here, but after the deregulation of the airline industry was put into effect in the late 1970s and early 1980s, Frontier got out. It had happened all over the United States—when airlines didn't have to serve smaller cities, when they weren't required to by the government, they cut their losses and departed for good.

So twice a day—three times on weekdays—the nineteen-seaters set down here; twice a day—three times on weekdays—they took off. I walked over to a machine selling big, various-colored gumballs, the old kind of gumballs in the old kind of vending machine you used to see in drugstores and on carnival midways—the sort where the gum rolled down clunky metal chutes with drop-down metal doors. A sign said that part of the proceeds would be "Donated to Civic Activities."

I tried to envision the railroad depot in its busiest years, with those thirty-two trains a day steaming right into downtown, letting passengers off, taking passengers on, both the depot and the passengers filled with the sensation of *being somewhere.* There was a series of clocks on the wall of the airport, each clock set to the time zone of a different city, and labeled as such. LOS ANGELES. DENVER. NORTH PLATTE. NEW YORK. LONDON.

I walked up to a window that was hot to the touch and looked at the runway in the sun, with not a plane to be seen.

"They said I was too valuable on the homefront," said John Zgud, eighty. "I was working at Martin Aircraft in Baltimore—I was the foreman in a sheet metal machine shop.

"My bosses told me that I was needed where I was, and that even if they did release me to go into the service, the Army wouldn't take me, because working in the aviation plant was too important to the war effort.

"But I said to myself, 'I'm going to quit my job.' I went to a tavern on the corner, and ate some crabcakes and drank some beer, and I thought about it.

"I knew I could stay a civilian, and have every excuse for doing so—I was helping to build airplanes. But when the war was over, people would have asked me, 'What did you do in the service?' And I would have had a new suit on, and money in the bank, and I could have said, 'I worked in an aircraft plant.' I was young. I didn't want to say that.

"When the war was over, I wanted to be able to say that I did my part."

So he quit the job in the aircraft factory, he joined the Army Air Corps, and he found himself on a train across the United States on his way to prepare for combat. He ended up as an aerial gunner on B-24s, flying thirty combat missions in Europe. But first was the train ride.

"It was a troop train made up of fifteen or twenty old passenger cars," he said. "We had to keep the windows closed, or all the soot would get in. The train was packed full—no shower, no place to eat. We ate these dried-up field rations."

It was one of those times in American life when the way in which people moved from place to place was irrevocably, everlastingly changing. He gave up a safe job in an aircraft plant . . . to ride on a train all across the United States . . . so that he could learn how to fly in airplanes that got shot at. It was his decision, and he has always believed it was the right one.

"After all that time on the train, and all those field rations, North Platte was almost too much to believe," he said. "These girls came out of the depot and toward the train with cigarettes, candy, chewing gum, telling us to go on inside and have some food and something to drink. Pretty, young girls—I still know a few of them."

That is because today he lives down the road from North Platte, in Cozad. He came back to the United States and got a job in Philadelphia as a plumber. He didn't much like it—and his wife said she had a sister who lived in a place they might enjoy more.

"Betty's sister lived in Nebraska," Mr. Zgud said. "We came out to see her, and we said, 'This looks like a pretty good place to settle.' And it has been."

When Mr. and Mrs. Zgud want to go somewhere, though, the highway is about the only option. The passenger trains don't come through here any longer—and the limitless promise of the air age is down to the thin drone of twin propellers, two or three times a day.

A sign on the wall of the airport building said that North Platte had been home to the nation's first lighted airfield. (The lights had consisted of fuel-burning barrels placed around the perimeter of the runway in 1921, so that a two-plane airmail caravan that was originating in San Francisco could touch down for servicing on its initial route east. One of the planes crashed soon after leaving California.)

There would be no more planes in North Platte this afternoon—no planes requiring lights, no planes not requiring lights—and I walked back into the parking lot, where the sun had turned the blacktop into goo that clung to the bottoms of my shoes. The sound of no noise was everywhere.

Whatever the birth of air travel may have promised the country—speed, convenience, the vanquishing of gravity—it took something away in the course of delivering those promises. What was taken away was, in no small measure, the country itself: the country as land, the country as place. On an overcast day, passenger jets don't just speed over North Platte—they don't even provide an opportunity to acknowledge it exists. People looking down from the sky have no idea it is there.

Those thirty-two trains a day, moving in and out of the old depot downtown—maybe there were days when the trains stopped, and not a single person got off, not a single person got on. Maybe the stops at North Platte, on days like those, were useless.

But at least the people on board knew they had tarried somewhere. At least they had to concede the fact of the town. If only by hearing the sound of the conductor's voice, announcing that they had arrived in a place that had a name.

"We lived in Lexington, sixty miles east of North Platte, in the war years," said Maxine Yost, eighty-two. "I was a young mother of two small children and, with the consent of my husband, I lined up a baby-sitter for the day. I volunteered to come with the ladies of the Presbyterian church to help in the Canteen for that one day."

If the North Platte airport was a reminder of how the way of arriving in town has been altered, Mrs. Yost's tale was a different kind of reminder—a reminder of how coming to town, even in the Canteen years, could hold hidden letdowns.

"We left Lexington early in the morning, bringing crates of hard-boiled eggs furnished by many of the nearby farmers," she said. "I was delegated to stay in the kitchen of the Canteen to peel the eggs, and it took many hours. I was disappointed to have to stay in the kitchen and not be able to talk to or visit with any of the servicemen."

That's what she had not expected: to work in the Canteen, yet never speak to a soldier. "I could only peek out once in a while and see the servicemen," she told me. "I wished I could work out in the dining room, and serve some of the men. But somebody else took the eggs out to the main room. I just wanted to wish the boys good luck—to tell them I hoped they had good luck where they were going, and that they would come back safe."

Mrs. Yost said that she does not regret having traveled to North Platte that day, although it was not what she had anticipated. "I knew that I was helping," she said, "if only by peeling the eggs for the boys to eat."

She was back home in Lexington in time to make supper for her husband, she said, and that night, when she thought to herself about what had happened at the Canteen, "I knew it had not been a wasted day."

Still, she said, it was a big world out there into which the young soldiers were being sent, and she has for a long time been sorry about one thing:

"I wish I would have at least walked out onto the platform. To see them get onto the trains and pull away. To see that with my own eyes, and to tell them goodbye."

Eighteen

At dinner one night, while waiting for my meal to arrive and reading some old news clippings I had found from the town's wartime years, I heard something: the *William Tell* Overture.

I didn't even have to look up to know where it was coming from, but I did anyway.

A cell phone—the phone of one of the other people eating in the restaurant.

It was a thin, trilling, reedy version of the melody—the *William Tell* Overture as it might sound on a tiny kazoo—which the man had programmed his phone to play instead of ringing in a conventional way.

He let it keep ringing—the man allowed the *William Tell* Overture to play again and again, as he checked the phone's display screen for the caller ID of whoever had placed the call. Then he could decide whether to take the call—while the rest of us in the vicinity of his table had to listen to the phone's music.

What it made me think of was not the usual assortment of complaints about how telephone discourtesy has taken over the land. Instead—because everything I was encountering in North Platte was in the context of the Canteen, and what had happened there—the phone and its musical ring made me consider another fact of life faced by all those young men on the trains, in the days before instant communication became an American birthright.

They went years without hearing the voices of the people they loved, some of those soldiers did—years without hearing the sounds they would have cherished. Now we expect that voices be delivered to us anywhere, any time— we demand it, really. The *William Tell* Overture beckoned the man who could not decide whether to permit the caller's voice into his life right now, and I tried to imagine it: riding across the country on a troop train, knowing the voices you adore will be lost to you for years, knowing you will not even be able to count for certain on receiving a letter from those who care the most about you. Cast adrift on your way to war, with no lifeline home.

"If someone wanted to mail a letter, you'd put a stamp on it for them."

Leona Martens, seventy-three, was telling me what would happen in those first moments when the soldiers hurried into the Canteen. She was a teenager in Wellfleet, Nebraska, during the war: "Just a jump and a holler," as she described the town of fewer than one hundred residents.

But then she would go to the Canteen.

"I was really young at first," she said. "A freshman in high school. Those boys would come into the Canteen like you'd known them for a long time. 'Well, would you like to see a picture of my girlfriend?' 'Sure I would.' They'd pull a picture out and show me.

"It was probably pride, and loneliness, too. They were proud to show you their girl—show you what they had. But they knew they were leaving that behind."

Her job inside the depot, she said, was preparing the fixings for sandwiches. "I helped with the meat grinder. It was a big commercial grinder that you ran by hand. It wasn't something they'd let the little kids do."

They wouldn't? I thought she had said she was young when she came to the Canteen.

"I was thirteen," she said "That's not little—not when you're born on the farm."

So she ground the meat—"it went further when you

ground it, put in a little pickle, a little mayonnaise"—and did whatever else was requested of her: "Anything they needed. One day I poured iced tea all day."

But the main thing was the boys—the young soldiers and sailors. "Any woman who tells you that flirting didn't go on isn't telling you the truth," Mrs. Martens said. "They forget what they were like when they were fourteen and fifteen and sixteen if they tell you they didn't flirt."

The era, of course, was quite different when it came to the ramifications of flirting: "Girls didn't go to bed with boys the first night. You just smiled and talked with them and maybe rolled your eyes a little.

"The night before, I would talk to my mother about what a fun day it was going to be. We would bake cookies and my mom would tell me that we were going to get around early in the day."

Get around early in the day?

"That meant get up right away and get breakfast over with," Mrs. Martens said. "And when we got to the Canteen, it was just kind of a rush. They had big double doors looking out on the tracks, and those doors would open up and the soldiers would come in off the train in droves. There'd be no one in there, and the first thing you knew, the place was full.

"I'd always be wearing a skirt and blouse. No pants on girls back then—we wore skirts and felt like pretty girls. When those boys came in . . . well. I've always been a flirt.

Don't let the other women you've been talking to fool you. These were handsome, eligible guys on those trains."

I asked her what the soldiers would say to her.

"You'd always hear, 'You remind me of my sister,'" she said. "Don't you know the lines boys use on girls that age?

"But a lot of times we'd just get done with one train, and here comes another one. You'd not hardly recover from the last train—you would have had fun with that bunch, joshed with them—and the next bunch comes in. A lot of them had been on their train from clear across the country.

"City and farm kids both, never been away from home, some of them. You could tell those—they were the ones who were so quiet."

Those were the boys, she said, who sometimes would have letters in their hands. They would have written home while on the train, and they would be looking around for a place to mail the letters, and the girls at the Canteen— having seen the looks in the eyes of lonely boys many times before—would walk up carrying postage stamps, and would take the envelopes, and would assure the boys that the letters would be on their way.

"There were always the boys who had the gift of gab," Mrs. Martens said. "They're the ones who had a circle of people around them. But then you would see the boys who didn't—the boys who didn't seem to know how to talk to anyone.

"Those are the ones you would walk over and say hello to. With boys like that, you really felt needed."

The American soldiers around the world may have been waiting for letters from home—letters that took weeks or months to arrive—but some of those soldiers across the oceans were writing letters addressed to North Platte.

As I read some of those old letters, I was struck by the courtesy, almost courtliness, of the young men who took the time to write. One of the letters, sent during the war to the *Telegraph* by a Private John L. Lewis:

Dear Sirs:

I am addressing this letter to you, a newspaper, in the hope that you will contact those parties of whom I write.

I am one of the contingent of soldiers who, while enroute from the West Coast, was treated to a display of thoughtfulness and unselfishness to warm any man's heart.

At the railroad station of North Platte, on or about the fifth of August, the troop train was met by a group of ladies armed with large baskets of smoking tobacco, oranges and candy which was given to the men in uniform.

I, as a member of those troops, have been delegated to write a letter of thanks and appreciation to the ladies and the city of North Platte. That act of courtesy and kindness will long be remembered by those men.

*In closing may I state, for some of us the English lan-
guage is far inadequate to fully express our gratitude. Hoping
this message reaches those responsible parties, I remain,
Sincerely yours,
Pvt. John L. Lewis*

And always—always—there were the letters from the
soldiers' mothers. What had surprised me at first—that the
mothers of fighting men who passed through North Platte
would expend the effort to put their thanks on paper and
mail them to the North Platte volunteers—was now
something I was accustomed to. The soldiers would write
to their mothers, and say how they had been treated at the
depot; the mothers would write thank-you notes to the
Canteen, as if not to do so would be unforgivable.

One such letter:

*I wish it was possible for me to step into the same place
that my darling boy was invited and treated so kind, so I
could tell you how much I, as well as he, appreciated it. Words
cannot express my gratitude to you.*

*I had a letter from him yesterday and he told me of the
great work you are doing. He said it was the first nice thing
that had been done for them. He said "Mother, you don't
know what it means to a fellow so far away from home."*

*Heaven bless you for what you did for my boy and every
other mother's boy—I am a widow and my right hand is crip-*

pled as you can see by my writing, but through the help of God I am trying to do all I can. Again I thank you for what you are doing. I will always remember you when I approach the Throne of Grace.

Some of the soldiers and sailors weren't as new to the ways of the wider world—they'd been around. Ralph Steetle, now eighty-nine, a retired public television executive who lives in Waldport, Oregon, was a Navy lieutenant junior grade and thirty years old when his train took him through North Platte.

"I had left my wife and daughter in Baton Rouge, and was on my way to San Francisco," he said. "It was a regular passenger train, and there were about a hundred of us military on board, and maybe three hundred other passengers.

"As we approached this one community the conductor yelled, 'All uniformed personnel out, next stop.' We thought, 'Oh, hell, someone's going to count us again.' Another checkpoint. We're going to be checked to see if we're still living, or whatever.

"As we get out, it's quite obvious it's not a military check stop. All these people come out of the train station, with things in their hands. It's a town project. I got hold of the local newspapers while we were there. The society news wasn't about parties or dances—it was about how many hams were taken to the train station, how many pies were baked.

"It's soon evident to us what they have done in this town. There had been some moments during the war—you might be at a bar in Chicago and someone would say, 'No serviceman here buys his own drink.' But nothing like this—nothing like what we found in this town.

"Inside the train station was a counter that must have run a couple hundred feet, and a white tablecloth covering it. It was loaded with all kinds of food, and the people made it obvious it was their gift. To us. You knew as soon as you walked into the room that you would never forget this.

"They had a birthday cake they wanted to give to anyone, but it was no one's birthday. One of the ladies finally said to this Army captain: 'Captain, it's your birthday.' And he said, 'All right,' and he took it back onto the train and we all ate a little of it.

"We'd all had a weak cup of coffee somewhere, but we'd never had a feast. I figured they were patriotic souls who wanted to do their part. They thought it up, and they did it."

He left San Francisco for the Philippines and New Guinea: "My last sight was the Golden Gate Bridge, and I thought, 'I wonder if I'll ever see this again.'" Occasionally, while on duty in the Pacific, he would recall those few minutes in the town where the people had been waiting. "There are certain times during the war, when you're going from tedium to nausea and back again, when you think of good things," he said. "You would remember North Platte, and you'd think, 'I wonder if those people are still doing it. If they are, bless their heart.'"

The delivery of the mail to the fighting men in the Pacific was sporadic at best. Knowing that communication would be difficult and much delayed didn't make the reality of it any easier.

"I had a child born while I was gone," Mr. Steetle said. "That was a solitary feeling, being away from my wife and knowing that the baby was coming.

"It was six weeks after my daughter was born before I knew we had a baby girl, and my wife knew that I knew. All those weeks, not knowing. That's just how it worked. Our baby's not a baby now. She's fifty-eight. Her name is Janie."

One cold winter day at the Canteen—according to a newspaper report of the time, which I came upon in the back of a cabinet drawer in town—the women volunteers encountered an ex-serviceman.

The former soldier was sixteen years old.

He was wearing no uniform—just light clothing with a pair of Army shoes.

The women had to persuade him to have something to eat. The story reported:

Living in a little town out of Chicago, he had enlisted. . . . His outfit was sent to San Francisco. Military authorities there questioned his age. He finally admitted he was 16 and received his honorable discharge.

He had then hitch-hiked his way from San Francisco and was resting in the depot when found by women of the Canteen.

The women of the Canteen called Hinman's [a local service station/garage] and related the story, and a transport truck was found which took the youth to Omaha, with the promise that a ride from Omaha into Chicago would also be found for him.

The boy did not depart, however, until the women had taken up a collection amounting to $7, which they presented the youth.

He could not thank them, as he broke down and cried. . . .

Nineteen

With all the warmth and good feelings of the Canteen years, it was potentially easy to push aside the thought that things could be more than a little dour and chilly back in that America. I found a remnant of that other America in the most unlikely of places: a display memorializing the medical profession of west-central Nebraska, in a corner of the county's historical museum.

Certainly the display must not have been intended to elicit somber and starched reactions; undoubtedly the people who put it together (and it appeared to have been in place for many years) meant for it merely to be an artifact of the era from which it came. But to spend time in that section of the museum, to realize that it represented a very real and daily part of North Platte during the Canteen years, was to understand yet another way in which our world has changed—in this case, not necessarily for the worse.

It started with the framed photographic portraits of the physicians who served the town and county back then. They may well have been friendly, compassionate, welcoming men (and they were, in fact, all men). But to look at those portraits was to be reminded of the American age in which the medical profession often seemed purposely distant, didactic, unalterably aloof. When doctors, at least many of them, appeared to pride themselves on ruling from on high, with little room for long discussion or questioning. Doctor knows best.

It was a time when doctors seemed to have been encouraged to cultivate an aura of mystery—an aura that made some patients vaguely afraid, or at least apprehensive. The portraits on the museum walls, almost without exception, showed austere men with stringent, even-lipped, not-to-be-challenged expressions; shot at the Brown Harano photographic studios in North Platte, the portraits presented formal-to-the-point-of-frostiness men in eyeglasses and business suits and a visible demeanor that translated to: Don't ask.

It may have been unfair to judge them by the photographs, taken all those years ago, yet there was no question that the portraits had been composed to set a tone a universe away from soothing. These were doctors as the remote, elevated, mirthless older brothers of the Wizard of Oz—men behind curtains not meant to be fully pulled back, ever.

And if the men weren't really that way, plainly it was how they wanted the world to see them—at least the portion of the world that arrived expectantly each morning in the waiting rooms of their offices in this part of Nebraska.

"I was very surprised to receive the letter," said Naomi Wood, seventy-two. "I received it in 1945, and I still remember opening it and trying to figure out how they got my name and address."

The letter was from a hospital—McGuire General Hospital in Richmond, Virginia. The letter had been written by the wife of Lieutenant Colonel Garnet P. Francis, a serviceman who had been rendered blind by a Japanese shell in the Philippines. He had been freed from a prisoner-of-war camp after being held captive for three years. Four days after he was set free, the Japanese bombardment struck him.

His wife, Earleen, was an Army nurse who also had been a prisoner of war. She helped her husband survive the shelling, and after they arrived back in the United States and were heading for the hospital in Virginia for treatment to try to bring back his sight, their train stopped briefly in North Platte.

That was what the letter was about. Evidently Lieutenant Colonel Francis had been given a cake at the depot—and evidently Naomi Wood (who at the time was sixteen-year-old Naomi Smith) had baked it, and had placed a card with it that contained her name and address.

"I probably put a birthday card with the cake, because we always assumed someone would eat the cake for his birthday," Mrs. Wood told me.

So, mailed from the hospital on the East Coast, there was the handwritten letter from the blinded soldier's wife. It arrived at the home Naomi Smith shared with her parents in Big Springs, Nebraska, near the Colorado border. Lieutenant Colonel Francis's wife began the letter:

"I am so sorry I have neglected to write, but until now I simply haven't had time. . . . We found the card [with the cake], which was so sweet. . . . He is in the Dental Corps. He has a chance to regain his sight in a year, maybe before. This is why we are here in this hospital. He is being treated and I'm staying with him."

Naomi Smith Wood told me she didn't recall baking that particular cake, but said she had always been eager to do whatever was needed during her Canteen duty.

"There was a group of ladies in Big Springs who would go to North Platte to help out—I forget whether it was the Methodist Club or the Happy Hour Club—and I liked going with them," she said. "You didn't always have an assignment when you got to the Canteen—you just let them know what you wanted to do, and they usually let you do it."

The man who would become her husband, Bruce Wood, an Army Air Corps pilot of B-24 Liberators, also grew up in Nebraska, the son of a farmer, and he had been to North Platte before the war: "The town was somewhat of a shopping center for us, and they had some medical facilities there." As a young boy he had been in the train depot, but when he went through as a serviceman, "They had rearranged things to accommodate the Canteen. They had scrambled to use up the space.

"There was a limit of time you could be in there and be served and be back on the train. If you were on a seventeen-car train and you were near the end of the train, you had a pretty long distance to cover to get to the Canteen, and get back. That's why we ran—to get there on time. If you dawdled too long, you were going to be stuck there."

Lieutenant Colonel Francis had not been one of the men who ran to the depot—in all likelihood, because of his loss of vision, the cake was brought to him on board the train. Naomi Wood has kept the letter from the hospital in Virginia for all these years. The wife of the blinded soldier wrote to the sixteen-year-old girl:

"Your cake was simply delicious. . . . After being a prisoner of the Japs for three years, and after being starved the way we were, it really tasted good. I hope I learn to make cake as good as yours when I start to cook. No one knows when that will be."

I told Mrs. Wood how remarkable I thought the whole thing was, and she seemed surprised that I thought it was so exceptional. Those kinds of things happened—at least they happened at the North Platte Canteen.

"I was very happy to get the note," she told me, "and I appreciated it." At the end of the letter from the hospital, Lieutenant Colonel Francis's wife gave an address in St. Petersburg, Florida, where they hoped to be living after he was released from treatment and accorded time to recuperate.

"I'm sure I sent him a get-well card," Mrs. Wood said. "And I hope he did. I hope he got well."

The medical equipment on display at the county museum—the array of tools used by the doctors of the war years and the years before—was as cold and forbidding as the expressions on some of the physicians in the portrait frames. The equipment was designed for healing, for making a human feel better; in these surroundings, though, laid out and labeled as if they were artwork, the devices made you want to turn your head.

Perhaps that was it—perhaps it was the setting that was jarring, not the instruments themselves. But to see them, to read their descriptions—the "long curved forceps," the metal appliance meant "for irrigating wounds," the dangerous-looking cutter "to remove adenoids" . . . to look at all of this was to gaze into a world the doctors lived in every day. No laser surgery or micro-incisions for them, not back then—these steel implements were their stock-in-trade; for many surgical procedures, they probably still are for the physicians of today.

To see the utensils in the display case near North Platte's Cody Avenue was to understand, just a little, the chasm there has always been between doctor and patient. The instruments here wouldn't cause a sideways glance, among surgeons. And the surgeons, after enough years in the profession, might not comprehend what a powerful effect the sight of the instruments might have on outsiders.

Not just the surgical tools, either, but the more mundane items doctors of the era carried around to house calls with them, the way housepainters carried cans and brushes. The little leather packet containing (as the label informed) an "ear and nose light"; the zippered "ear, nose, throat" case, with the examining probes inside in need of sterilization after each use; the hypodermic needles in their own boxes, apparently also reusable and not to be tossed away; the big, hollow glass tubes to feed medicine into those needles, before being boiled and made ready to accept more medication for more needles . . .

The metal pincers to assist in the extraction of internal

organs, the long surgical scissors, as functional and harshly utilitarian as a paid-by-the-round prizefighter . . .

Prairie medicine, preserved in this catch-you-off-guard setting.

"When the hospital trains would come through—those were the days that would get to you."

Doris Kugler, eighty-three, was telling me about the toughest moments for the Canteen volunteers. She said a feeling of sadness came over the place when the trains carrying wounded soldiers would steam into North Platte.

"We were not allowed on the hospital trains," she said. "We were only permitted to walk up to the cars, and hand up the baskets of sandwiches and apples and oranges and candy and cookies. Whatever soldier was able to walk to the end of the car was the one to whom we would hand the baskets.

"Sometimes you could see the guys looking out the windows of the train. You would see bandages. Some of these young kids were on stretchers, and seeing them, so injured . . .

"And of course, you would see all of this knowing that your husband was overseas fighting. . . ."

Mrs. Kugler and her husband, Bill, had resided in McCook, Nebraska, south of North Platte, where he had been the manager of a shoe store called Pat's Bootery. While he was in the Navy, Mrs. Kugler lived in North Platte with her mother, who had come to stay with her.

"We got along fine," Mrs. Kugler said. "My mom got a job in a ready-to-wear store called the Mars Shop, on Dewey, and also got a job at O'Connor's Department Store. My dad had passed away in 1940—he was only fifty-six—so my mom said she would stay with me while Bill was in the service.

"A bunch of us wives got together and we formed a club. We called it the War Wives Club. Eight or ten women—we would go out to eat at Tucker's Restaurant on Jeffers. We all worked, and none of us had children, so we would have dinner and maybe a drink before we ate.

"The soldiers would come over from the military base in McCook, and sometimes they would get a little fresh. The owner, Mr. Tucker, would say to the soldiers, 'These women are all my friends, and they're all married—stay away from them.'

"When I think about it now, the soldiers were probably lonely. They probably just wanted some ladies to talk to. We were all happily married, so nothing was going to happen—we thought it was kind of funny when Mr. Tucker would keep them away from us."

She said that most of the members of the War Wives Club made scrapbooks for their husbands who were overseas: "News items, stories about parties, things from

Halloween—the idea was to save souvenirs from the time our husbands were gone, to give to them when they got home.

"I wrote Bill every day for two and a half years, from 1943 to 1945. I heard a lot of news around town. People would tell me things, and I would write it in letters to him. He was on Saipan, in the Marianas, and I think he got every letter I wrote."

I asked her if she ever skipped a day—with weekends, she certainly could have gotten away with doubling up some days and taking other days off from writing.

"I could have done it that way, but I didn't," she said. "Writing every day was something I wanted to do."

Her husband came home in 1945. "I went to Omaha and met his train," she said. "He got off . . . it was kind of strange. He always looked young in his Navy uniform. He didn't look hardly different at all. We got back in the groove pretty quickly."

Her mother insisted on moving out so that the husband and wife could have their privacy; Bill Kugler got his job back at Pat's Bootery, where he worked for the next thirty-three years, until his retirement in 1978.

As for the War Wives Club, they all stayed married to their husbands. "We were pretty dedicated," Mrs. Kugler said. "One of my friends from the club is ninety-one now. The others, I believe, are all gone."

Mr. Kugler died in 1985; he was seventy-one, and he had been ill for more than five years.

"He had twelve operations," Mrs. Kugler said. "One of the surgeons said to me one day, 'Bill must have a lot of money, because you're sure stickin' to him.' You get to know the surgeons pretty well during a long illness like that, and you become friendly and you try to make each other smile.

"I told the surgeon, 'He doesn't have money, but I'm stickin' to him.' And I'm very glad I did. I'm glad I stuck to him."

There was an iron lung in the museum, beneath the row of doctors' portraits. From one end of the machine extended the head of a dummy, constructed to resemble a young woman.

The iron lung—made of solid gray-green metal—was, according to the manufacturer's identification label on its side, a product of the Wiesner-Rapp Company Inc., of Buffalo, New York. During the national outbreak of polio, when iron lungs were hailed as the one hope for life for young sufferers, this particular machine evidently was the first one purchased in this part of Nebraska, and was placed for use in St. Mary's Hospital.

The girl in the iron lung—the dummy—was meant to

represent a specific, identifiable individual. The dummy's face stared up toward the ceiling—her neck was encased in rubber so that the pressure inside the chamber could be maintained; there was a gauge on the side to measure that very pressure. The dummy was intended to portray an eighteen-year-old woman from Cozad—she was named in the display—who had been the first patient from the area to be put into the lung.

From a news account, posted as part of the exhibit:

AN EIGHTEEN-YEAR-OLD VICTIM OF INFANTILE PARALYSIS WAS PLACED IN THE LUNG LAST NIGHT AT ABOUT 9:30 O'CLOCK AND REMAINED IN IT THIS AFTERNOON.

Then an update: The young woman was "improved." Then "slightly improved."

Then:

THE ENTIRE CITY WAS GRIEVED ON TUESDAY AFTERNOON WHEN IT WAS LEARNED THAT [here the young woman was identified by name] HAD PASSED AWAY AT ST. MARY'S HOSPITAL IN NORTH PLATTE. A WIDE CIRCLE OF FRIENDS WILL MOURN HER PASSING, AND ALL WILL EXTEND HEARTFELT SYMPATHY TO THE BEREAVED PAR-ENTS AND OTHER RELATIVES OF THIS FINE YOUNG WOMAN.

Such a peculiar exhibit; such an uneasy feeling. I looked at the head extending from the iron lung, immobilized; I looked at the glistening metal instruments in the nearby display cases; I looked at the impassive faces of the doctors in the picture frames above.

It was difficult to conceive what it must have been like to be the young woman inside the machine—the first person in North Platte to be sealed in the pressurized lung, her every change in condition reported to a curious public reading the daily news. Whether she wanted strangers to know or not.

I looked again at the face, at the eyes gazing straight up. There was one word to describe the patient: helpless.

Twenty

"Knees up! Knees up!"

The expanse of grass I came upon while walking through North Platte early one morning contained a tableau as far from the sickroom gloominess of the museum exhibit as anything could be on a luminificent summer day—as removed from the darkness of youthful illness and infirmity as any scene you could ever encounter.

"Linebackers! Over here!"

Summer football practice—the first sessions, preparing for the fall season. These were North Platte High School athletes, but practice was being held on the field behind

Adams Middle School—these boys weren't sure things to make the varsity, they were younger students, many of them trying out for the high school team for the first time, and they did not rate the main football field, at least not yet.

I wandered over to watch, and stood with two of the coaches, men in shorts and T-shirts with whistles clipped to lanyards. "We're just taking a look at them," one of the coaches told me. "Seeing who we have here."

And in the eyes of the young football players you could tell that they were wanting to be seen, hopeful to be noticed. Entering high school in the middle of Nebraska, figuring out what it means to be a young man, aspiring to be thought of highly, to be admired . . . you could see the boys sneaking glances at the coaches, trying to perceive if the coaches might be looking back at their section of the field. Each boy seemed to be attempting to run a little faster than the one next to him, shout a little louder during calisthenics, be a little quicker to his feet after the stretches were completed. Anything to be singled out.

"All right!" a coach yelled. "Everybody down to this end of the field!" They sprinted, while at the same time trying not to appear too overly eager, and it all translated, as it forever has at summer football tryouts, to: I hope the men in charge will take note of me, and approve.

"Linebackers!" That word again, hollered again, and you could tell that for some of the boys the word was tantalizing, a promise, something they dreamed would soon enough officially describe them. To be a linebacker, on the North Platte Bulldogs varsity on Friday nights in autumn . . .

How do you break out of the crowd? That is what this all came down to: a chance to break out, maybe for the first time in your life; a chance to be special, a chance, in this high school, to be a part of the group at the top of the hierarchy.

The coaches circumspectly pointed out to me the boys they thought had the best prospects for making the team, and some of the boys saw the coaches pointing, and I thought about boys not much older than these, boys who once were taught, at basic training, that breaking out of the crowd was not supposed to be their goal. Boys who rode the trains into and out of this town, many years before this summer morning, members of a team with different kinds of uniforms. "For our away games, we travel as far as Lincoln," one of the coaches told me, and I thought about young men who were on traveling squads that hurried through here on their way to Europe, on their way to the Pacific. Away games.

"It's difficult for me to think about the war without thinking about my high school football classmates," said Anthony J. Barak, seventy-nine. "I somehow always find myself thinking about what happened to them."

He grew up in Petersburg, Nebraska, went to Cathedral High School, and worked for a year to save enough money to go to Creighton University in Omaha. But in 1943 the Navy sent him to midshipmen's school in Illinois: "There was a great need. We had a double class going at the same time."

In Fort Pierce, Florida, "We had amphibious boat training—the floating coffins, they called them. We practiced our landings. Then we were supposed to go to Coronado, California. The troop train was a mixture of Army and Navy guys.

"That troop train . . . it was like a cage full of monkeys. You couldn't rest, you couldn't sleep—there was just seating, no beds. It was cold outside as we rolled through Nebraska, but it was so hot in the train. We were supposed to sleep sitting up in our seats or lying down in the aisle.

"It was going to be three days from Omaha to San Diego. At North Platte we kind of went off on a side track—that was the first inkling I had that something unusual was going on. There was a little girl standing there by the side of the train. We got off, and I went up to her and took this package from her. It was a cake. I was so surprised, I kissed her on the cheek.

"Although I had grown up in Nebraska, all I really knew about the town was that it was in sandhill country. My dad was a federal banking examiner, and he traveled through many towns. He had always told me that North

Platte was a railroad town, full of railroad people and cow-boys from the ranches.

"So that was my expectation. But what happened there . . . we were there for, at best, twenty minutes. We were treated like that not anywhere else the rest of the trip. No one else did anything like that for us. They were mostly older ladies at the North Platte station, and they made us feel sort of like it was our mothers, greeting us and hugging us. It was like being home, a little. Home for twenty minutes."

Home would soon seem like the most distant of mem-ories. "I was at Okinawa," Mr. Barak said. "Our ship was in the picket line off Okinawa—we were supposed to pick up any incoming kamikaze aircraft or submarines.

"I often think about the trip over there. It was really our first experience with rough seas. You got awful seasick. The only real training for something like that we had was on Lake Michigan, which was very choppy—but the Pacific Ocean, those huge waves . . .

"They could slap you either left or right. Sometimes the ship would heave and pitch. It dived down into the waves. If you were on the deck, and the ship would surge, you saw the horizon go up and down. I lost the fight a few times. I got so sick.

"It's the most horrible thing, and there's nothing you can do about it. You should have smelled the troop quar-ters on the ship, where we were housed. You'd go to the men's room, and you'd slip on vomit on your way there. People were getting sick all the time. It was a pretty awful business, I'll tell you. I would rather hit a beach than sit on a troop ship and be so sick."

When he had left Nebraska for the war, "I had a girl-friend back home. She gave me up for being lost. She thought I was going to come back in a coffin. Because of that, I felt that people didn't really appreciate what we were doing for them. Because of *her.* But then I would think about North Platte, and about people who *did* care about us."

He returned from the service and studied assiduously, earning a Ph.D. in chemistry. He spent his career teaching biochemistry and internal medicine at the University of Nebraska Medical Center. He has been married for fifty years; he and his wife raised five children.

When he thinks of his time in the war, it is not necessarily his shipmates who fill his memories. He recalls young men who were not at sea with him. Mr. Barak—like Paul Gardner, who had told me about his lost fellow high school football players—remembers the teammates from his school.

"I played three years for the Cathedral football team," he said. "We were the Cathedral Cardinals—red and white. We were really pretty good.

"I can see them now. I lost one in India, one in New Guinea. You'd hear about it when it happened. You'd get the news that some of the guys from your high school football team were getting killed.

"Hooks Hamilton, our halfback. Herman Jadlowski, who played tackle. You're a kid, and you played next to them on your high school football field . . . and then they're dead."

The North Platte football team continued their summer workouts and, as the sun rose higher in the sky and the morning heated up, some of the players took their shirts off and tossed them to the side of the field.

I found myself smiling as they did it, because a few days earlier I had met Gene Slattery. The ultimate take-off-your-shirt kid.

He's sixty-eight now, a farmer—but when he was nine, at the beginning of the war, he devised his own way of making money for the Canteen.

"It started with some goats," he said. "We lived over in Big Springs, and my dad took me to a livestock sale, and I was helping to sell some goats down on the floor and someone who knew my family kidded me by yelling: 'Why don't you sell your shirt, too, Gene?'

"So I did—I took it off and sold it. I gave the money to the North Platte Canteen, and people talked about it.

"I got the idea to keep doing it. Every livestock sale I could get to, I would go down onto the floor of the sale barn and the auctioneer would auction my shirt off. I did it all over this part of the state.

"People started buying it and then giving it back—they'd pay for it and say, 'Let's sell it again.' So I would sell it many times at the same livestock auction. The buyers knew the money was going to the Canteen.

"And whenever someone did keep the shirt, there were clothing-store owners in North Platte who told me: 'Now, Gene, any time you need a shirt, you come to me.' They would give me shirts for free, to sell at the auctions."

He did it for four years: "The most I ever got for a shirt was seventeen hundred dollars, at a War Bonds drive at the Paramount Theater, across from the Pawnee Hotel. Right up on the stage."

Then the war ended, and the Canteen closed, and there was no more reason for him to do it. He was thirteen; soon he started high school, and did all the things that high school boys in the late 1940s did, including playing on the football team. He was a lineman.

"But no one knew me as a football player," he said. "Everyone knew me as the kid who sold his shirt."

"When I graduated from high school in Sutherland, there were only fourteen students left in my class, and we had started as freshmen with thirty-three," said Marge

Roethemeyer, seventy-three. "Only three of our graduates were boys. The rest . . . well, you know. The war."

The boys in her class had been sent off to fight, leaving high school life behind. When she was fifteen, she began to travel the twenty miles to North Platte with "the older women" of Sutherland—"they seemed old to me at the time, but thinking back on it now, many of them were young married women"—and would volunteer to help at the Canteen.

"I told the ladies in North Platte that I was sixteen, although I wasn't yet, because they had that rule that if you weren't sixteen you couldn't work on the platform, and I wanted to. I was kind of a tomboy, and I liked being outdoors.

"My dad had been in World War I, and he had been gassed and shell-shocked. He seemed always to be sick and always in veterans hospitals. He was gone a lot when we were kids.

"Living with my dad being like that all those years, I just did what I could. When World War II started I was a freshman in high school, and I helped the Red Cross fold bandages. That was before pre-made bandages—they didn't have things like we have now. You had to fold and cut the bandages by hand.

"The Canteen always felt like a home to me. It was as homey as could be. There wasn't room for everyone to dance, but some soldier would always want to. He'd say 'Let's dance'—he would probably be seventeen or eighteen years old—and somebody would play the piano, and we'd do the jitterbug.

"It would never last very long—maybe five minutes. That's not very long anyway, but especially when the boys were going to be there for such a few minutes. You'd hear someone call 'All aboard,' and all the boys would tear out and get on the train. You'd be kind of downcast that it would be ending so quickly.

"Some of those boys would make an impression on you at first light. A couple of them, I wrote to. A lot were going overseas. They would ask, as they were leaving the Canteen, 'Will you write to me?' They weren't flirting. They were lonely.

"Sometimes it was really cold out on the platform. You would stand out there and wave at the train as it pulled away. They would be looking at you through the train windows, and you would wave goodbye until it was gone."

The North Platte football players were running wind sprints, sweating in the sun.

I talked with the coaches a while more, and then headed off to try to find something I had been hearing about. My time in town was growing short; I would be leaving soon. But after thinking so much about the trains that didn't bring people here anymore, I wanted to see a certain piece of land to the west and the north of the main city. I wanted to learn if what I had been told could possibly be true.

Twenty-one

"You can't get back here on your own."

Deloyt Young, the retired manager of operations at the Union Pacific's Bailey Yard, was behind the wheel of his car. He had arranged to meet me in the parking lot of a convenience store out on a public road, and now he was taking me past checkpoints and guarded gates. We were still on the outskirts of North Platte, but it certainly didn't feel like it—this was a part of town that only the people who work inside its private fences regularly see.

"They turn people away who try to come in," Young said. "This is a place for working, not for looking."

It was astonishing—every foot of it. And in this town,

the town where the passenger trains stopped coming all those years ago, the existence of this place . . .

Think of it this way: What if O'Hare International Airport was set down here in North Platte, and no one ever talked about it? What if LAX was here, or Hartsfield?

That was the Bailey Yard. Ask a thousand people on the street, anywhere in the United States, what it is, and they are likely not to have the slightest idea. Unless they are railroad people.

The Bailey Yard—in North Platte—is the biggest railroad yard in the world.

In the town the passenger trains left behind.

It runs twenty-four hours a day, every day of the year. It covers 2,850 acres—and stretches eight miles in length. The people who work there have figured out that if the University of Nebraska football team were to choose to play its games in the Bailey Yard, there would be room for 3,097 football fields.

What happens at the Bailey Yard is at the same time quite elementary, and almost impossibly complex. Freight trains heading east and west arrive here, pulling cars with cargo destined for different cities. But not all of those freight cars are supposed to end up where the locomotive is going to end up. One car on the train might have cargo that is supposed to be in Pittsburgh, one car on the train might have cargo that is supposed to be in St. Louis, one car on the train might have cargo that is supposed to be in Detroit. . . .

The locomotives can't just drop the freight cars off in the individual cities, like children in an after-school car-pool. The railroads don't work like that. Classification yards are needed—places where the freight cars can be quickly and efficiently pulled off the trains they arrived on, and added to trains that will get them to where they are scheduled to end up. It's like shuffling a deck of the biggest and heaviest cards in the world—and the shuffling never stops.

"Come with me," Deloyt Young said, easing his automobile up to the base of a tower that looked like something you might find at a major airport.

We climbed some steep metal stairs high into the tower, until we were looking out a large window. It was a humbling sight.

Railroad tracks everywhere—endless, jammed with trains from all parts of the United States, more trains than you have ever seen all at one time. A flurry of activity on every track—time was what mattered here, getting the cars switched and attached to the proper trains was the business of the Bailey Yard.

"We've got fifty westbound tracks and sixty-four east-bound tracks," Young told me. "The man who controls all the traffic is called the humpmaster. There are three hundred fifteen miles of track in here. Every twenty-four hours, this yard handles twelve thousand railroad cars."

I couldn't even see to the extremities of all the trains. This town that had been a part of America's greatest love affair with passenger railroad trains, this town that had given birth to the Canteen, this town that was unreachable by passenger railroad today . . . it was home to the biggest working railroad yard anywhere on the planet. With not a single paying passenger on any of those twelve thousand cars.

"It's simple," Young told me. "They figured it out a long time ago. They can make more money off freight. It takes about half as many people to haul freight as it does to haul people. And the people all want to fly, anyway. They don't have time to ride the trains. At least they think they don't."

But once they had no choice. Once, they didn't even know exactly where they were going.

"We weren't told," said Edward J. Fouss, eight-one, who now lives in Eufala, Oklahoma. "We didn't know our destination. Not in those days. Loose lips sink ships."

The Navy was sending him to the South Pacific in 1944, but had not informed him of that fact. All he knew was that he was on a troop train that he had been ordered to board in New York, that was traveling west, and that held, by his estimate, between 1,500 and 2,000 men.

"The bunks were stacked high," Mr. Fouss said. "There was one cooking car on the train, and you ate using your mess kit and your canteen. You would carry that to the cooking car, and they would ladle out your food for you. Beans, and I don't remember what else. Potatoes, and what passed for coffee."

He laughed and said, "It wasn't very romantic."

For some of the men, it seemed the trip would never end: "It took five days and five nights to get to California. When we were in Iowa going west, they said that in the next state there was a place where we were going to be allowed off the train, and that we would be given something to eat. We didn't know what to believe—it didn't sound right. We were *never* allowed off the train, so what was this about?"

The men found out soon enough. "All the tables, all the counters, all those wonderful ladies," Mr. Fouss said. "Anything we wanted to eat . . . I never, in all these years since, have figured out why they were the ones to do this. In the whole country, why them?

"The first thing you saw in North Platte was that it was very *orderly.* Just a very nice feeling, the way everything was set up. And it *smelled* so good in that town. The bread . . . well, Nebraska is in the middle of the United States, so I guess it shouldn't be surprising that they have the right crops to make bread that smells good, and it did. We didn't get off the train again until we arrived in California.

"Maybe being in the Middle West was a big part of why it could happen there. The people in the Middle West are a different breed of cat. They don't live with their heads in the clouds. They are willing to think things are OK. They'll give you a hand."

When he came home from the war he set up a radio shop, doing both sales and service. Then television came along, and he worked putting electrical wiring into houses. The world was changing, and he knew it—more and more, people stayed home and let the world be delivered to them on a glass screen instead of venturing out quite as much to try to see that world with their own eyes.

And when they did venture beyond the borders of their towns, as often as not it was high in the air. "When you're flying, you don't see much," he said. "When you traveled by train, you could see the little towns—you could see America."

That's how he saw North Platte—and even though it was only for a few minutes, he has never forgotten.

He has no souvenirs from his brief time in North Platte, no photographs, but that doesn't matter. "I still see the town," he said. "What I see, when I think back on it, is a lot of happy people."

He never would have met them if the troops had been

transported by airplanes. And after the war, that's the way just about everyone in the United States began to travel.

"The railroads decided they could make more money hauling a hog than carrying a person," Mr. Fouss said. "The railroads did it to themselves."

At the Bailey Yard, Deloyt Young steered his car past rusty tanks and around outbuildings with grit and streaked dirt on their windows, his wheels bouncing over ten sets of tracks until we arrived at an enormous shed inside of which Union Pacific mechanics repair and service more than ten thousand locomotives a month. He led me inside, and once we were there he had to shout to be heard over the work being carried out by hard-hat crews on every side of us.

"These guys can work on eighty-seven locomotives at a time," he said, leaning close to my ear. "It never stops— three shifts every day. The railroad keeps a hundred rooms at a motel a few miles from here—they're turned over two or three times a day from crews coming in, getting some rest while their trains are being serviced here, then leaving town while other trains and other crews come in."

The yellow locomotives with the red, white and blue Union Pacific logos painted on them dominated our sight line, but Young motioned to locomotives and freight cars owned by other railroads, too, all being worked on here. Men were climbing over and under and around the cars, and if the men needed to be reminded of the dangers of being too casual around these huge machines, there was a toteboard keeping track of how well the employees were meeting their safety goals during a given period of time. Things appeared to be going relatively well: 0 COLLISIONS. 0 AMPUTATIONS. 0 FATALITIES.

Young pointed out to me, beyond the repair shed, some different kinds of trains rolling in and rolling out: "That's a double-stack train leaving town, heading east. That one next to it is a soda-ash train. And you can see that coal train on the next track."

He said that Americans don't think much about freight trains, even though much of the merchandise and goods they couldn't live without is delivered to them over the rails. "I don't know," he said. "People felt they were a part of the passenger trains. They don't feel a part of freight, and I doubt that they really think about how everything they need gets to them."

Off in the distance—it didn't seem to be emanating from inside the Bailey Yard, but the place was so large that you really couldn't tell—came the sound of a train whistle.

"Two longs, a short, and a long," Young said. "He's getting ready to go through a crossing."

"To understand what the passenger trains meant to North Platte, you have to be from there," said Fran Hahler

Wohlpart, sixty-five, who was born in North Platte and who now lives in Las Vegas.

Right before World War II, she said, "people made their own entertainment. On Saturday nights, you'd have supper, get in the car, and go downtown. Your dad would get a good parking space, and you would watch the people go shopping—including a lot of the farmers who would always come to town to shop on Saturday nights.

"You just sat there in the car and said, 'Oh, there's Louie—he looks well this week. He's got his whole family with him.'

"The passenger train station was the center of downtown. I'd be sitting listening to my mom and dad talking, with my brother and my two sisters in the car with us—my father was the General Motors dealer in North Platte. Hahler Buick—he sold Buicks, Cadillacs, Olds, Pontiacs. . . .

"And we would sit in his car and watch the people. This was before television, so that was the only way to see something—to be there. You would watch the people walk from store to store—O'Connor's Department Store, Rhoads Dress Shop, Montgomery Ward . . . this is what a lot of families did on Saturday nights."

Once the war started and the troop trains began rolling through, the scene at the railroad station downtown became electrifying, Mrs. Wohlpart said. "The thousands of uniforms pouring out of the trains—Army, Navy, I was too young to always know what each uniform meant, but

I knew they were all on our side. The thing I remember— and this is not what you might expect—is that most of the time *they all seemed in such good humor.* It's as if they were eager to get where they were going, that they were looking forward to what their country had asked them to do.

"The Canteen at the train station was so crowded, all the time. I didn't have any business in there, but I loved watching from the outside. The trains were always coming in and going out—there was never a quiet hour. If coming downtown to watch people used to be a Saturday night thing, once the Canteen opened it was an every night thing. My parents would see a soldier and say to us children, 'He's a nice-looking guy, isn't he? I think he'll really do well for himself in the Army.'

"Looking at all the soldiers at the train station inspired us. My mother rolled bandages, to be used on the battle-fields. My aunts knitted sweaters, vests, scarves, socks and mittens. How did it get to the soldiers? How did anything get to the soldiers—you turned it over to the government, and they got it to the soldiers."

And those nights at the depot, just watching the loco-motives come in and out of town, with America's sons on board . . .

"You would be surprised how many people did it," Mrs. Wohlpart said. "I was there hundreds of times. I'd go in the afternoons with my mom. There was a constant stream of soldiers getting on and off the trains.

"Mom always wore her corset and her stockings and her shoes with a medium heel, and a dress, of course, because this is the way ladies dressed. And always a hat—you didn't leave the house without a hat. In the summertime straw, in the wintertime felt."

I asked her if, had television been a part of America's life during the 1930s and 1940s, she thought the people still would have come downtown—not only during the Canteen years, but during the years before and immediately after the war, when families came to the train station and the stores to watch everyone out on the town.

"No," she said. "People wouldn't have come downtown if they had television sets, because there would be something to watch at home. TV's more convenient. You don't have to do anything."

There was no shortage of things to do in the Canteen years: "Even as a kid, I remember having very busy days during the war. We had paper drives, rubber drives, scrap metal drives—my sister and I would take a little red wagon around, and once on a scrap metal drive we hit the jackpot. Someone gave us a water heater. I was only six, and my sister was four years older. You'd just knock on the door. 'Hi, we're on a scrap metal drive.' They'd give you cans, or a pot. And then one day there was this water heater.

"They helped us put it on the wagon. We were pulling the wagon, trying to balance it—it seemed like fifty miles to us, we were so little. I wish we could have seen ourselves. We were thrilled. By a water heater."

She has been back to downtown North Platte, in the years after the passenger trains stopped coming. Their absence, she said, changed everything.

"Dead," she said. "Downtown just felt dead without the trains. The trains *were* North Platte. Without the passenger trains, there wouldn't be a North Platte. The town came into being because of the railroads. If the Union Pacific had laid its tracks fifty miles to the north, that's where the town would have been.

"The feeling of the town, now that the train station has been torn down, and now that there hasn't been a passenger train pull in or pull out in thirty years . . .

"It's hard to explain, unless you were there. The place was *alive*."

In the Bailey Yard, in the locomotive repair shed, a sign warned the laborers working on engines what could happen if they did not follow the safety precautions:

HANG YOUR FLAG AND HANG YOUR TAG
OR YOU COULD END UP IN A BODY BAG

Deloyt Young explained the wording to me—it referred to various signals the workers were required to post on trains in the shed to caution other crew members when a locomotive was under repair, with a human being crawling around it. Across the way I could see a freight train so long that it seemed to have no beginning or no end. I asked him about it.

"There's eighty cars per mile," he said. "A one-hundred-eighty-car train is two and a quarter miles long."

And trains that long really move across the country?

"They do," he said. "Big coal trains can haul thirty million pounds of coal. On one train."

Trains miles long, with not a soul aboard, save for those driving the locomotives. I asked Young how big the crews were for these giant trains.

"Two," he said.

"Two people?" I asked, thinking I must have misheard him.

"These are very efficient trains," he said. "It only takes a crew of two to operate them."

All those train cars, all the way down the tracks. The biggest railroad yard in the world—and not a passenger to be found.

It was dimly lit in the locomotive shed. I would be leaving North Platte in the morning, but there was a place I wanted to see one more time.

The place where the depot used to be—the ghost of the Canteen. The downtown train station that wasn't there. The home of all that life.

Twenty-two

Late in the afternoon, with sundown on its way, it was quiet next to the double sets of tracks on Front Street.

This was it—the place of no people, save for the men still drinking alcohol straight from the bottle near the shade of the short brick wall and planter that commemorated what used to be.

The people who worked downtown would be starting home for the night, if they hadn't already. They most likely would not be passing by here—this part of Front was a shortcut at best, not a destination.

I walked over until I could see the metal of the tracks.

This had to be it, I calculated; this had to be exactly where the platform of the depot had stood.

The headline atop page one of the North Platte *Daily Bulletin*—the edition of Wednesday, August 15, 1945—consisted of letters so tall and so thick that the men in the composing room almost certainly had to have crafted those capital letters by hand. They were once-in-a-lifetime letters, not of the sort routinely kept in type for daily use.

WAR ENDS!

The words stretched the width of the page, designed to be seen from across a city street. The *Bulletin* is gone now, as are the days of hot type in newspaper shops, but on that day in 1945, both did their jobs quite well. The banner headline was like a joyous voice screaming in the town square.

The story beneath, transmitted by United Press, began: "President Truman has announced the government of Japan has accepted the Potsdam ultimatum without any qualifications. Simultaneously, from the White House and 10 Downing Street in London has come word that the world, after nearly six years of war, again is at peace."

Directly under the main story, a version of which you

could read in any newspaper in the United States, was something the specifics of which you could get only in North Platte: the local reaction to the end of World War II.

The headline was considerably smaller, although still three banks high: CITY CELEBRATES V-J DAY; CHURCHES OPEN FOR PRAYER; MANY PLANS PENDING. The story, written by Larry Hayes of the *Bulletin* staff, started off:

> A few minutes after 6 o'clock, North Platte promptly went mad. Residents poured into the loop district as whistles shrilled, bells rang and horns squawked.
>
> Business houses began shooing patrons out as soon as the United Press flash became known. Streets were filled with motorists and pedestrians alike. . . . Traffic was jammed. Firecrackers popped, smiles and waving hands were on all sides. . . . Impromptu snake parades were formed about the streets. . . .

But variations of that story, too, were appearing in papers in every city in the country—stories of the celebrations on hometown streets. Only the names of the streets and the names of the towns were different.

In North Platte—lower still on the front page of the *Bulletin,* with a one-column headline—was a story that could have been written nowhere else.

CANTEEN HEARS V-J NEWS, the heading read.
And the report:

> Every describable emotion was expressed at the
> Canteen last night when news of the Japanese surren-
> der reached the visiting service men. Some cried, most
> of them just shouted, and still others stood numbed,
> unable to believe what they were hearing. Meanwhile,
> the Canteen board, realizing the importance of contin-
> uing to greet the returning veterans with smiles and
> treats in the months to come, made plans for the future.
>
> The War Dads of Kearney, forty-five strong, were
> on hand with a large donation of supplies. . . .

That was the news from Front Street that August day—
the war was ended, but the volunteers from Kearney had
driven a truck over with 480 candy bars, magazines, 10
crates of oranges, 80 cases of soft drinks, 400 loaves of
bread, 300 pounds of meat, 3,000 hard-boiled eggs, 75
sheet cakes. . . .

It was as if they didn't know that everything had just
changed—or perhaps it was that they could hardly absorb it.

Within a day, the news in town had already shifted to the
next phase—to plans for a city, and a nation, at peace.

The headline above the story by Larry Hayes the day after was: NORTH PLATTE TO RE-OPEN TODAY; VICTORY DANCE FOR CANTEEN SLATED TONIGHT. It began by informing the towns-people:

> Stores will be open this morning, and it will be "business as usual" in North Platte.
>
> After the first wild outburst of rejoicing, the city assumed a calmer aspect yesterday and observed peace for the first time since Pearl Harbor, with a complete closing of all retail business.
>
> Gasoline rationing went out and blue points are no longer a worry. . . . Hotels were turning patrons away as tourists sought to lay over. . . .
>
> The biggest local attraction for the V-J celebration will be the big Victory dance at Jeffers pavilion tonight. The dance will climax the city's celebration, and, fittingly, will be a Canteen benefit affair. The Royal Nebraskans will furnish the music, and the dance is being sponsored by the North Platte Lions Club.

Local merchants had hastily put together display advertisements promoting not their products or services, but the end of the combat. Munson's Texaco Service Station, 1020 North Jeffers, sponsored an ad showing a drawing of a cemetery field filled with crosses, with the heading: A LASTING PEACE——THAT THOSE DEAD SHALL NOT HAVE DIED IN VAIN.

The Fairmont Creamery Company, 112 West Sixth, took out an ad with a small drawing of a dove carrying a ribbon in its beak, alongside the words: PEACE AGAIN. TODAY, WHEREVER THE FOUR WINDS BLOW, THE EYES OF MANKIND BEHOLD THE VISION OF THE BETTER WORLD TO BE.

The Canteen, while announcing that its job was not finished, began to acknowledge that the end was coming:

> The North Platte Canteen will continue to operate up to ten months following victory day, Mrs. Adam Christ revealed yesterday. Mrs. Christ, chairman of the general committee, said, "We will still have a tremendous task before us and I'm sure the people in the surrounding territory will continue to give us full support."
>
> She reported there would be a true need for the center as the millions of personnel in the armed forces return. "We can fill a mission of real welcome to the returning veterans," she said.

I stood in the emptiness where the Union Pacific station had been. I thought about whether the Canteen volunteers could ever have contemplated this—could have looked ahead and seen in their minds a day when the building would be vanished, when no trains would stop on Front Street.

I had spoken with a woman—Ann Perlinger, sixty-nine, of Paxton—who had been one of the youngest Canteen volunteers, and who had seen the place go from being a dream in her excited eyes, to being a vacant lot she now made a point not to visit.

"I was nine years old when I began to help out," she told me. "We didn't get to do much—we put the popcorn balls together. We just popped the corn and made a syrup—sugary water, really. We formed the popcorn balls and wrapped them in waxed paper.

"They were put into the baskets that were carried to the troop trains. One day I guess I decided to put my name and address inside the wrapper.

"I received a letter from a soldier—a Private William Washille. Paxton only had about seven hundred people, and here was this letter from a soldier in the Army, arriving at our home.

"Inside the envelope was a little note from him saying he had come through North Platte, and had seen my name in a popcorn ball. I wrote him back and said, 'You're going to be disappointed—I'm only nine years old.' He then sent me a Christmas card."

Everything about the Canteen captivated her: "I remember the milk bottles. Pint bottles and half-pint bottles—the soldiers just tipped them up to their mouths and drained the milk right out of the bottle, then put the empty bottles back." Her mother was the postmaster in

Paxton, she said, and "I was always going to the post office, because we had a box there. I was allowed to open it. I would get all excited—I told my mom, 'I'm hearin' from my popcorn ball.'"

She grew up and got a job as a secretary to Paxton's school superintendent, and then a job at Paxton's only bank; she met a man and married him, and they farmed until his death in 1998. They had five children, two of whom now live in Paxton, two in Omaha, and one in Council Bluffs, Iowa.

She often travels the thirty-two miles from Paxton to North Platte—"We have no shopping here in Paxton"— and every time she pulls into town, "I can remember it like it was yesterday.

"All those young men, so many of them, all those uniforms . . . I guess it made the world seem big. We didn't know where those boys were from or where they were going. When they would leave, I would want to get on the train with them.

"When I go to North Platte now, I never go down by the railroad tracks. There's not much down there anymore. All those boys . . . they were here and then they were gone."

For eight months after the end of the war, the Canteen stayed open, still there for every train that came through town. The soldiers and sailors were coming home, and they were rolling across the country again, this time heading toward peace.

So the volunteers continued to report to the Canteen, but after a while it did not feel quite the same. There was not the urgency; there was not the sense of being so needed. As the spring of 1946 approached, almost all of the soldiers who were returning home were home.

The decision was made. The announcement was carried in a brief story in the newspaper:

> April 1 was set as the definite date for closing the North Platte Canteen at a meeting of the general Canteen workers Tuesday night. The vote was almost unanimous. . . .
>
> Letters will be sent to the towns and committees who have supported the Canteen, informing them of the closing date. They will be invited to choose a date to serve before April if they wish to do so.

On April 1, 1946, sixteen trains were scheduled to bring soldiers through North Platte on the Canteen's concluding day of operation. Mrs. Charles Hutchens, secretary of the Canteen at the end, put out a request "asking that all persons who have loaned cooking utensils and other equipment for the duration of the Canteen call for the articles tonight."

The Ladies Aid of St. John's Lutheran Church of Gothenburg was the main group of volunteers that final day; they had been scheduled for the day, so there they were.

The last soldier to enter and then leave the Canteen— the last of the six million—was Charles H. Plander, of Marshalltown, Iowa. He and eleven other servicemen had arrived on a train that was making the customary brief stop; they did not realize that the Canteen had just closed for good, and encountered three volunteers doing last-minute cleanup. They were Mrs. Hutchens, Mrs. T.J. Neid and Mrs. Amiel Traub.

"Don't feel bad about closing the Canteen," Charles Plander said to the women. "You've earned enough points for your honorable discharge."

Mrs. Hutchens, Mrs. Neid and Mrs. Traub, their work completed, had just finished making a large pot of coffee for themselves.

They gave it to the soldiers instead.

Twenty-three

It was getting toward dark, and I knew I should be leaving the railroad tracks—walking away, for the last time, from where the Canteen had been.

I was finding it difficult to do. There was nothing here—yet I still wanted to stay a while more.

I thought about some of the people from the Canteen years, men and women with whom I had spoken earlier. Two of them in particular—at the end of our conversations, they had told me what the town had felt like to them after the Canteen was gone, and after the depot had been torn down.

Doris Dotson—the woman who had collected the

patches from the soldiers, the woman who, when she was a girl, had made the insignia jacket for herself; the woman who, because of her stroke, was no longer able to dance— had told me this before we said goodbye:

"It was very sad when the Canteen stopped. We still had troops on trains going through, but not very many of them, and the people on board soon enough found out that there was no Canteen.

"It was such a good feeling that the war was over—a good feeling for the whole country—but here in North Platte there was a sense that suddenly there was a void. You just kind of felt lost."

Larry McWilliams—the man who had told me that North Platte was once "a rough and wide-open town," a city that felt "like a little Chicago" before the Canteen made it wholesome—had said this to me at the end of our talk:

"After the Canteen closed in 1946, it was a period of postwar wind-down in North Platte. The passenger trains were still coming through, but there were fewer and fewer soldiers on board.

"Still, it was fascinating to watch. In the late 1940s, there were still big differences in this country. The way people talked, the way people dressed. Passengers would get off the trains at the downtown depot, if just to stretch their legs before getting back on, and it was *polyglot*. The different ways the people spoke . . . East Coast people getting off the trains, more stylish than you would see in central Nebraska . . .

"It's more difficult to distinguish different accents these days, even in a place like Texas. I think television has brought the country into one language, one sound, now. You can still hear accents, but not as much.

"In central Nebraska, for instance, *film* was 'fill-um.' *Arthritis* was 'arthur-itus.' *Get* was 'git.' *Wash* was 'warsh.' You could stand at the train station and listen to the people who got off, and you really could tell that they were from someplace else.

"When I come back to North Platte now, I go downtown, and I think about what it felt like when the streets were always packed, and the train depot was always filled with people. It almost doesn't register for me—I look at it, and it almost doesn't register that the depot isn't there."

Within two weeks of the Canteen's last day of operation, its volunteer officers had typed up a final audit that was as painstakingly fastidious and devoted to exacting particulars as had been the Canteen itself.

Dated April 13, 1946, the audit reported that the bank balance as of March 1 had been $4,181.22, and that the balance at the close of operations was $3,033.56. Expenses during that last month had included $766.29 for sugar and fruit, $19 for eggs, and $21.70 for flowers for retiring

Union Pacific Railroad president William F. Jeffers, who had donated the space at the train station for the Canteen.

The audit noted that the first deposit in the Canteen's bank account, on December 27, 1941, had been $56.76: "Only during the first month or so of operation did Canteen finances bring headaches to the management; then the generous and patriotic people of western Nebraska and eastern Colorado learned of the commendable and satisfactory service that was rendered to the uniformed men and women passing through, and from then on voluntary cash contributions monthly were sufficient to pay all bills promptly."

The audit conceded that so much food was donated to the Canteen by the people of the area that no accurate accounting was possible: "The value of the food contributed during the operation of the Canteen is a mere matter of guessing."

The Canteen's pocket-size deposit book, issued by the First National Bank of North Platte, showed that at least once a week for every week of the war someone from the Canteen visited the bank to put a few dollars in, which was duly noted by a teller's hand on each page of the ledger. Whenever there was a withdrawal—such as on May 8, 1943, when $294.75 was taken out for expenses, dropping the account to $257.74—a deposit was promptly made, in this case on May 11, taking the account back up to $354.50.

Every detail was looked after at the end. The members of the Canteen's steering committee gathered for a group photo before the furniture and equipment were removed from that part of the train depot; the committee voted not to use any of the remaining funds for the purchase of a plaque on the side of the train station to memorialize the Canteen. The women of the board decided that "a memorial of some sort is appropriate," but that "inasmuch as the funds were given for the benefit of servicemen," any expenditures for a plaque "should be raised in some other way than taking them from the Canteen balance."

The committee voted that the $3,033.56 on hand at the close of the Canteen should be used for the benefit of veterans in hospitals. Equipment, including recreational items for the recuperating soldiers and sailors, was purchased and donated to the Veterans Administration Hospital in Lincoln, the state capital. One hundred thousand matchbooks were purchased, to be distributed at regular intervals to the hospitals in Lincoln and in Denver; on the covers of the matchbooks was printed: *With the Compliments of the North Platte Canteen.*

On October 1, 1946, John R. Yorby, chief of special services for the VA hospital in Lincoln, wrote to Mrs. Charles Hutchens, the secretary of the Canteen board in North Platte: "It is impossible to express adequately our appreciation for the many donations which you have made to the hospital. . . . After working with you grand people, we can understand why the Canteen was such a grand success and your praises sung by service people all over the country. . . . We hope that the friendship and good will between us will continue on through the years."

"I was working as a tool-and-die maker in Kenosha, Wisconsin," said Harry Mulholland, seventy-nine, who still lives in the Wisconsin town of West Allis. "I was twenty years old. This was 1942. I quit my job to go into the service. I was overweight for the Coast Guard, but the Army didn't care. I was a body."

He ended up spending twenty-six months of his thirty-eight months in the service overseas, including at Utah Beach. He stopped at the North Platte Canteen three times—in 1942 heading east to west, then twice in 1943, once heading west to east and once heading back.

"In those days, the trains stopped for water and coal," he said. "On that first trip, we were added on to a regular passenger train. You were with the other guys in the troop cars—the unknown made me nervous. I was a twenty-year-old kid, the first time away from home. . . . You didn't know what was going to happen, or where you were going to go.

"You would sit on the siding during some parts of the trip west, as other trains would go by. There were a lot of cornfields—not much to see. You slept in a berth with some guy you never saw before. No shower—you could shave, but that was about it. You felt filthy.

"As we were getting close to a town in Nebraska, someone who had been through there said, 'Wait till you see this.' I don't know why, but I believed him.

"And we arrived in North Platte, and . . ."

He waited a few seconds, and then, through his tears, said:

". . . the kindness . . ."

He had to stop again.

"I'm sorry," he said, struggling. "Being away from home—my mom and dad and my brother and I, we were such a close family, and being away from home . . ."

He said that after the train left North Platte, "Everyone in the troop cars was so enthusiastic about what had just happened—how wonderful it was. And the next time I came through, I was able to tell the guys on the train, 'Wait till we get to North Platte.' "

On that second trip across the country, Mr. Mulholland said, "It was a mob scene at the train station, so many people—it had been not quite a year since my first time at North Platte, but what was happening at the train station seemed to have gotten bigger. Just as friendly, though. The idea that the people had done this—the whole time I had been in the service, that they had been doing it every day and every night at their train station . . ."

His third time through, he said, "I was thinking that

overseas was coming up. Again, the unknown. On that train, you slept wherever you could sit. And when I left North Platte that time, I didn't know that I would not see them again, but I thought that might be the case, and it was.

"They were the greatest doggone people. . . ."

He and his wife of forty-five years live in a retirement community now. He told me that "many years after the war, we were on vacation, and we had driven across the country to California. On the way back to Wisconsin, I decided that I wanted to have a look. I wanted to go to North Platte."

He had told his wife all about the Canteen, of course. "We got to the edge of town, and there was a restaurant near the highway—a Perkins or something. We went in to have lunch.

"There was a young waitress. I asked her if she knew anything about the Canteen. She said she thought her folks might have worked there.

"I said to her, 'You thank your folks for me.' She was eighteen or nineteen. There wasn't much there. She was hardly listening to what I was saying. She was born after the war. I could tell that what I was saying didn't mean anything to her.

"My wife and I finished our meal and left. We just got back on the highway. I've always thought, in the years since, that I should have gone into the Chamber of Commerce or something that day. Found someone to thank.

I'm pretty sure the waitress never passed along our thanks to her parents. That's all I wanted to do when I was in town—to thank somebody."

I left the place where the train station used to be, taking one last look. I was going to have some dinner at the hotel, then pack so that I could leave town in the morning.

Around eleven o'clock that night I was feeling restless and couldn't sleep, and I remembered that in the hotel's restaurant there had been some old photographs of North Platte through the years, including photos from the war years. I had seen them as I'd eaten that evening, and on other evenings during my stay.

I thought I'd go back downstairs and spend just a little more time with the pictures. When I got to the restaurant, though, it was closed for the night, its lights off. But the adjacent bar was open, if not exactly crowded; the bartender said that I was welcome to turn the restaurant lights on and wander around looking at the photographs.

So I did. There was a photo of the women of the Canteen serving the soldiers at one of the long tables covered with dishes of donated food; there was a photo of the soldiers and sailors running inside from a train; there was a photo of a workman adjusting the Canteen sign outside the depot as the volunteers stood watching.

And then there was one I hadn't noticed before. It was

a photograph of North Platte on V-J Day—it looked as if it had been taken in the evening. The downtown streets, in the area of the Pawnee Hotel, were filled with cars, mostly black sedans. It was a celebration—downtown was jammed bumper-to-bumper—but if a big celebration can ever be described as orderly, this one appeared to be. The war was over; the people of the town had turned out to rejoice together.

Tonight, all these summers later, in the bar of my hotel the television sets were showing a sports event to the few people still having drinks, but the play-by-play was turned down, and music from a tape or a jukebox filled the room. I could hear it—a song by the rock group Kansas.

I looked at the faces of the men and women in the photographs—the young men in uniform stepping off the troop trains, the women working in the Canteen—and, as the sound drifted over from the barroom, I listened to the words from the song.

Dust in the wind,
All we are is dust in the wind. . . .

I stayed a few more minutes, then departed for my room, turning off the lights on my way out.

Acknowledgments

There are many people who helped me during the writing of this book. I'd like to thank them here.

At William Morrow, the professionalism on deadline of Tom Dupree and his assistant, Yung Kim, was a source of steady reliability. Michael Morrison, Lisa Gallagher and Sharyn Rosenblum have made my association with Morrow a pleasurable one.

At Janklow & Nesbit, Eric Simonoff, once again, was instrumental to my work with his advice, counsel and friendship. To Mort Janklow, as always, goes my appreciation and admiration.

In Nebraska, the fine assistance of Katie Klabusich was

invaluable. Katie's diligence, her eagerness to work long hours, and her eye and ear for detail were a constant help as I did my reporting and research.

At the *Chicago Tribune,* my thanks for making it possible to write my daily newspaper columns while working on this book, as well as for all the kindnesses over the years, go to Ann Marie Lipinksi, Howard Tyner, Jim O'Shea, Gerould Kern, Dale Cohen, Jim Warren, George de Lama, Don Wycliff, Bruce Dold, Tim Bannon, Geoff Brown, Mary Elson, Jeff Lyon, Scott Powers, Mike Zajakowski, Marsha Peters, Chris Rauser, Bill Hageman, Nadia Cowen, Bill O'Connell, Stacy Deibler, Stephan Benzkofer, Rhonda Rogers, Digby Solomon, Ben Estes, Jill Blackman, Charlie Meyerson and Dan Kening. Off the newsroom floors, my thanks go to John Madigan, Dennis FitzSimons, Jack Fuller, Scott Smith, Owen Youngman and David Williams. At WGN television, I am grateful to colleagues Carol Fowler, Steve Sanders, Allison Payne, Al Romero, David Parrish, Jennifer Lyons, Mitch Locin, John Owens, Eric Scott, Brad Piper and Vince Rango.

It was North Platte resident Cal Robinson who, many years ago, first told me about the Canteen, and about what had happened in his town during World War II. Mr. Robinson's emotion for, and pride in, the place where he lived were so heartfelt and vivid that I never forgot what he had told me—and I always promised myself that one day I would get there to see and hear for myself. I thank him for that—and for helping me to understand the city and the Canteen.

For a reporter to go into a town as a stranger, and try to tell the town's story, would be a nearly impossible task without the generosity and goodwill of the people who live there. I cannot imagine any reporter being treated with more openness and kindness than I was by the people of North Platte. Many of their names you have read in this book, but so many more, through their taking of the time to speak with me, to share their memories of their town, and to make me feel at home were just as important in helping me tell the story of the Canteen. I will never be able to thank them enough. They are:

Marilyn Peterson, Marion Effenbeck, Lloyd Speicher, Dean Hiner, Faye Fisher, Gene Smith, Ken Killham, Cecil Jacobson, Walt Carlyle, Marge McHarness, Charles Bostwick, Karen Hipwell, Lori Schoenholz, Dick Linn, Bob Kukas, Bruce Friedrich, David Bernard-Stevens, Jackie Lashley, Muriel Barrett, Kathy Swain, Mike Sharkey, Laura Johnston, Christian Zavisca, Kelley Gavin, Tonya Cross, Pat Dannatt, Frank Graham, Jenny Hasenauer, Sharron Hollen, Eric Seacrest, Craig Jones, George Lauby, Mark Lewis, Kristi Nixon, Antone Oseka, Misty Ryun, Tad Stryker, Rick Windham, Derek Lippincott, Dan Fisher, Alicia Rivera, Barb Fear, Amanda Olson, Deb Brooks, Tonya Carroll, Chuck LaLanne, Andrid Olsen, Heather Shlaight, Jessica Simpson, Aaron Brannigan, LeeAnn Thompson, Lisa

Blessin, Shirley Woodruff, Tammy LaLanne, Mary Anderson, Sheila Snyder, Tim Anderson, Howard Webster, Neal Baxter, Marge Jones, Merilyn Mitchem, Milo Schavlik, Irene Bystrom, Frances Newberry, Joe Radzlaff, Nina Johnson, Dorothy Berryman, Les Weil, Anna Glaze, Jim Beveridge, Marilyn Brohman, Eunice Brown, Eileen Furrow and Ann Smith.

I would like to say a special word of thanks to Pasquale Grillo, an Army infantry soldier who passed through North Platte on a troop train in 1944, and who in recent years has suffered a stroke that makes conversation difficult for him. He spent long hours providing written answers to the questions I sent him, because he wanted to do everything he could to help with my reporting. Even though he can no longer talk without effort, his voice is very much a part of this book.

My admiration goes to two Nebraska historians whose affection for their part of the country, and their work in preserving its place in the nation's tapestry, not only provided valuable background information for me, but was of incalculable help in enabling me to locate the volunteers from the Canteen years, and the soldiers who once passed through North Platte. These historians are James J. Reisdorff and Martin Steinbeck; I hope they will feel I have done a worthy job in telling this story.

At the Union Pacific Railroad's headquarters in

Omaha, Mark Davis was gracious in allowing me to have access to the railroad's archives.

For the sections of the book dealing with the history of the sandhills, I owe particular thanks to Marianne Beel, who explained to me the significance and nuances of that part of Nebraska life.

More than anyone else, though, the person to whom I am grateful beyond words is Keith Blackledge, to whom, along with his wife, Mary Ann, this book is dedicated.

Mr. Blackledge turned seventy-five years old during the time I was reporting and writing *Once Upon a Town.* He is the retired editor of the North Platte *Telegraph;* he led that paper for twenty-five years, from 1967 to 1992.

His friendship toward me, his endless patience with my questions, his willingness to walk me through various aspects of North Platte's past . . . all were indispensable to my work. His love of, and uncompromisingly high standards for, his town are evident in everything he does.

Most impressive is his absolute sense of honor and fairness. He comes from the tradition of smaller-town editors who know that, unlike their big-city counterparts, they are likely to encounter on the streets each day at least one person about whom their paper has reported in that morning's edition—so they had better be confident enough in what they have printed to in good conscience look that person right in the face.

Acknowledgments

I made certain that the last stop I made before leaving North Platte was at Mr. Blackledge's house. I will say publicly here what I said in private to him that day: If every newspaperman and newspaperwoman approached the job in the clear-eyed and honest way that he does, with his readiness to give every person a chance for an unprejudiced hearing and just consideration, then we would never have to worry about the public having a low opinion of our business. My hope is that Mr. Blackledge's spirit is on every page of this book; it would not exist without him.

About the Author

Bob Greene is a syndicated columnist for the *Chicago Tribune;* his reports and commentary can be read daily at www.chicagotribune.com/greene. As a magazine writer he has been lead columnist for *Life* and *Esquire;* as a broadcast journalist he has served as contributing correspondent for *ABC News Nightline.* His news commentaries can be seen on television superstation WGN.

His bestselling books include *Duty: A Father, His Son, and the Man Who Won the War; Be True to Your School; Hang Time: Days and Dreams with Michael Jordan; Good Morning, Merry Sunshine;* and, with his sister, D. G. Fulford, *To Our Children's Children: Preserving Family Histories for Generations to Come.* His first novel, *All Summer Long,* has recently been published in a new paperback edition.

OCT 2 4 2002